SpringerBriefs in Computer Science

Series Editors

Stan Zdonik
Peng Ning
Shashi Shekhar
Jonathan Katz
Xindong Wu
Lakhmi C Jain
David Padua
Xuemin Shen
Borko Furht
V.S. Subrahmanian
Martial Hebert
Katsushi Ikeuchi
Bruno Siciliano

For further volumes:
http://www.springer.com/series/10028

Zhongming Zheng • Lin X. Cai • Xuemin Shen

Sustainable Wireless Networks

 Springer

Zhongming Zheng
Department of Electrical & Computer
  Engineering
University of Waterloo
Waterloo, Ontario
Canada

Xuemin Shen
Department of Electrical & Computer
  Engineering
University of Waterloo
Waterloo, Ontario
Canada

Lin X. Cai
Huawei Technologies Inc. Rolling
Meadows
Illinois
USA

ISSN 2191-5768   ISSN 2191-5776 (electronic)
ISBN 978-3-319-02468-4   ISBN 978-3-319-02469-1 (eBook)
DOI 10.1007/978-3-319-02469-1
Springer Cham Heidelberg New York Dordrecht London

Library of Congress Control Number: 2013950796

Printed on acid-free paper

Springer is part of Springer Science+Business Media (www.springer.com)

Preface

With the environmental consciousness and increasing of energy cost, efficient energy efficiency consumption has become one of the most important issues in wireless communication networks. The eco-friendly and renewable green energy, such as solar and wind energy, is emerging as a promising alternative energy source. It is anticipated that green energy will be widely adopted in next-generation wireless networks in order to sustain the ever-growing traffic demands, while mitigating the effects of increased energy consumption. Unlike traditional energy provided by electricity grid, green energy is replenished from nature and highly depends on the capacities and locations of the electronic devices. Thus, the fundamental design criterion in the network deployment and management is shifted from energy efficiency to energy sustainability due to the sustainable nature of green energy. In this brief, we focus on network planning, energy modeling and resource allocation by jointly considering cost and energy sustainability in wireless networks with sustainable energy. First, we present the characteristics of green energy and investigate existing energy-efficient green approaches for wireless networks with sustainable energy. Second, we study network planning and resource allocation issues based on statistical historic data in green wireless networks to minimize the cost and sustain network. We jointly consider the relay node placement and sub-carrier allocation (RNP-SA), and then formulate the network planning and resource allocation issues into a mixed integer non-linear programming problem. After that, two low-complexity heuristic algorithms, namely RNP-SA with top-down/bottom-up algorithms (RNP-SA t/b), are introduced to solve the non-linear programming problem in different network scenarios. Third, we try to address the random availability and capacity of the energy supply based on accurate energy harvesting and consumption information. Our goal is to maximize the energy sustainability of the network, or equivalently, to minimize the failure probability that the mesh access points (APs) deplete their energy and go out of service due to the unreliable energy supply, where the energy buffer of a mesh AP is modeled as a $G = G = 1 (= N)$ queue with arbitrary patterns of energy charging and discharging. Based on the analysis, a distributed admission control strategy is proposed to guarantee high resource utilization and to improve energy sustainability. Finally, we conclude the brief and provide future research directions. It is anticipated that this brief will provide valuable guidance on the design of future sustainable wireless networks.

Contents

Acronyms

QoS	Quality of Service
BS	Base Station
AP	Access Point
RN	Relay Node
PV	Photovoltaic
WLANs	Wireless Local Area Networks
SNR	Signal to Noise Ratio
p.d.f.	Probability Density Function
MEDP	Minimal Sum of the Energy Depletion Probability
CDF	Cumulative Distribution Function
ME	Minimum Energy
MPRT	Minimum Path Recovery Time
RNP-SA	Relay Node Placement and Sub-carrier Allocation
MINLP	Mixed Integer Non-linear Programming
RNP-SA-t	RNP-SA with top-down algorithm
RNP-SA-b	RNP-SA with bottom-up algorithm
STR	Sub-carrier and Traffic over Rate
LHS	Left Hand Side

Chapter 1
Introduction

The growing user demand and the expansion of wireless communications have led to a tremendous growth of energy consumption in wireless access. Due to the limited battery capacity of mobile devices and the increasing cost of energy from the electricity grid, energy efficiency has become one of the most essential research issues in wireless communications. Many studies have investigated how to minimize energy consumption to extend the network lifetime [1–3] or maximize energy efficiency or energy utilization [4–6] of a communication network powered by traditional energy.

With the increasing concern on environmental protection and the preservation of natural resources, green energy sources, which refer to the energy harvested from nature resources and can be replenished without compromising the energy requirement of future generations like solar, wind, tides, etc., are anticipated to be widely used in next-generation wireless networks. The sustainable and eco-friendly characteristics can help to fulfill the ever-increasing user demand, while reducing the detrimental effects of conventional energy production. To promote sustainable wireless networking, communication industry has launched a few projects, e.g., Green Wireless Communication by King Abdullah University of Science and Technology [7], and WindFi: Renewable-Energy Wireless Base Stations by Steepest Ascent Ltd. [8]. In the field of wireless communication networks, green wireless devices, i.e., wireless devices like base station (BS), access point (AP) and relay node (RN) powered by green energy as shown in Fig. 1.1, have been developed to replace the traditional wireless devices. It was reported that wireless backbone devices, such as BSs, consume over 80 % of total energy in communication networks [9]. The deployment of green wireless devices in wireless backbone can significantly save the energy consumption and sustain the network operations. Therefore, it is anticipated that green wireless devices will be widely deployed to construct sustainable wireless network, i.e., wireless networks powered by green energy. Nowadays, many companies are upgrading existing conventional devices with wireless devices powered by green energy. For instance, Huawei [10] has extensively deployed green energy solution to power BSs in China, Africa, and the Middle East region; Sprint sets the goal that 10 % of its total electricity will be supported by renewable sources by 2017 [11].

Z. Zheng et al., *Sustainable Wireless Networks,* SpringerBriefs in Computer Science, DOI 10.1007/978-3-319-02469-1_1, © The Author(s) 2013

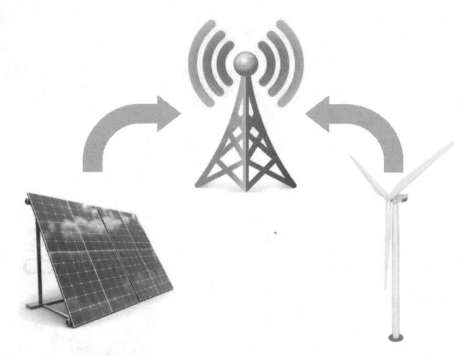

Fig. 1.1 An illustration of green BS

Different from traditional energy sources, green energy is usually considered as a sustainable resource in long term but its availability and capacity are dynamic and even intermittent in short term. For example, the energy generated by a wind turbine is dependent on the scales of wind strength, and the solar panel provides energy supply that varies across the time of a day and the season of the year. Due to the sustainable nature of green energy sources, the previous works on minimizing network energy consumption of traditional energy source from electricity grid are not applicable for the sustainable wireless networks. The main objective in sustainable wireless networks powered by green energy is no longer energy saving, but how to utilize the harvested energy to fulfill the traffic demands and to ensure network operations. Therefore, the fundamental design criterion and the main performance metric of sustainable wireless communication networks have shifted from energy efficiency to energy sustainability, i.e., the harvested energy can sustain the traffic demand of network users. There are many existing works focus on energy efficiency in wireless networks with traditional energy sources, while energy sustainability in sustainable wireless networks has not been well studied. Thus, we are motivated to revisit the research issues including energy modeling, resource allocation and network planning under the new dimension of energy sustainability in sustainable wireless networks [12, 14, 29] in this brief. We first present the characteristics of green energy supply. Then, we investigate how to provide qualified services and full connectivity to all wireless users by deploying relay nodes (RNs) and allocating sub-carriers into the network according to the historical statistics of energy harvesting and user demand information. To further improve the network performance in terms of energy sustainability, a generic analytical model is developed to study the energy buffer evolution based on the real time energy harvesting and traffic demand data. By characterizing dynamic energy charging and discharging processes, we can get closed-form solutions of energy buffer analysis. After that, we propose resource management and admission control schemes to mitigate the energy depletion probabilities of APs based on the analysis. In this chapter, we first give an overview of some research issues including network planning, the modeling of energy, and resource allocation schemes. Finally, we list our outline related to the research of sustainable wireless communication networks.

1.1 Network Planning

The network planning plays an important role in constructing wireless infrastructure to ensuring that a new telecommunications network or service meets the needs of the subscriber and operator. Generally, network infrastructure consists of diverse network devices and peripherals, such as BSs and APs, etc. As traditional wireless devices are powered by either batteries or fixed power, the deployment of traditional wireless devices is expensive and limited by various concerns, e.g., cable and battery cost. Basically, traditional network planning focuses on deploying wireless devices to either fulfill the QoS requirement of users [16, 17, 30] or minimize the energy consumption to extend the network lifetime [18–20, 31]. To construct sustainable wireless network infrastructure, the first step is to deploy green energy powered wireless devices into existing networks to replace wireless devices with traditional energy sources. Normally, in this step, we do not have detail energy harvesting and consumption information. To estimate the energy replenishment and traffic demand in a certain area, we can use the statistical historic data to approximate the energy charging and discharging capability. Based on the approximation of green energy harvesting and consumption, it is possible for green energy powered wireless devices to sustain the network operations and fulfill users' QoS requirement. However, green energy supply is highly environment dependent, thus some particular criteria should be included in the network deployment problem, e.g., local charging capability, traffic loads which reflect the energy demands on the device, and the available energy in the battery. For network planning problem in sustainable wireless network, it is essential to balance the energy harvesting capacities of each green wireless device with the traffic demand of users' requirement. Therefore,energy sustainability should be considered as one of the fundamental design criteria to guarantee that QoS requirement of users can be fulfilled by the harvested energy.

1.2   Energy Modeling

In traditional wireless networks powered by batteries, one of the fundamental issues in wireless communications is to improve the energy efficiency of the network due to the limited battery power of mobile terminals. The energy is usually considered as a limited but stable resource during the battery lifetime. Another form of energy can be obtained from the electricity grid, which can provide continuous and stable energy for wireless devices. However, the power is primary generated from limited and non-sustainable resources, such as coal, natural gas, and petroleum, thus this kind of energy is not sustainable. In sustainable wireless networks, the energy is regarded as sustainable but dynamic. To make the green energy usable, a rechargeable battery is used to store the harvested energy and output a relatively stable energy supply. Generally, there are two methods to alleviate the dynamic characteristic of green energy: Hardware design and energy modeling. Hardware design mainly focuses on how to stabilize the harvested energy to fulfill estimated energy requirement, which may be obtained from the approximation. There are many studies on hardware improvements, such as the photovoltaic (PV) power system sizing for solar panel [22–24]. In [22], authors propose a simple method for sizing stand-alone photovoltaic systems by using loss of load probability and load profile to represent the desirable reliability and traffic demand, respectively. In [24], various sizing tools for PV stand-alone system are analyzed. Moreover, a standard simulation model for PV sizing problem is designed. Comparing with hardware design, energy modeling mainly focus on balancing the energy harvest and energy consumption to stabilize the charging and discharging processes based on accurate energy harvesting and consumption information, which has the advantage of more accuracy, lower cost and easier to be implemented. The utmost critical issue of energy modeling for sustainable wireless networks is how to build recharging and discharging models. The energy modeling should be able to not only predict the remaining capability in the near future, but also help other resource allocation approaches to optimize the performance in sustainable wireless networks. For example, by comparing sustainability performance of various routing algorithms, authors in [34] show that traditional routing strategies without considering dynamic energy charging capability achieve poor performance, and it is crucial to adapt routing decisions to the time varying energy supplies. Therefore, it is essential to design an accurate analytical model of energy supply dynamics in sustainable wireless networks.

1.3   Resource Allocation

Usually, the infrastructure network resources are inherently limited while different customers have various QoS requirement with variable capacities in their customer devices. To provision satisfactory services to customers, efficient resource allocation is required to allocate the limited network resources based on the diverse requirement of users. Usually, the main objective of resource allocation in wireless networks with traditional energy sources is on load balancing, i.e., to evenly distribute the traffic load of wireless devices to achieve the maximal network throughput or the minimal delay, and to extend the network lifetime. In sustainable wireless communication networks, energy charging capability becomes the new dimension to be considered to balance the charging and discharging capabilities of green wireless device. Based on the analysis of energy modeling, we can estimate the energy harvesting and consumption to predict the sustainability of green wireless devices. Since different green wireless devices have various energy charging capabilities and traffic load, the resource allocation between green wireless devices becomes very important. For example, we can design routing strategy to allocate more traffic load to the green wireless devices with less energy burden/more redundant energy, such that green wireless devices with more energy burden/less redundant energy will not drain their energy soon. Therefore, it is critical to guarantee the energy sustainability by jointly considering energy charging and discharging capabilities in sustainable wireless networks. Some works [14, 26, 32] re-visit the resource allocation problem to achieve the network sustainability in sustainable wireless networks. In [33], power saving mechanism and control algorithm design are considered in solar-powered WLAN mesh networks. A statistical power saving mechanism and a control algorithm are proposed to maintain outage-free operations of the node match based on the future load conditions and solar insolation. In [12], a resource management scheme was proposed to distribute the traffic load across the network according to the dynamic energy charging and discharging processes.

1.4   Outline

Resource management for sustainable wireless communication networks is still embryonic. Due to the particular characteristics of green energy, the fundamental design criterion and the main performance metric of sustainable wireless communication networks are shifted from energy efficiency to energy sustainability.

Under the new sustainable networking paradigm, many fundamental research issues need to be re-visited. Instead of minimizing the energy consumption, the main goal of research management in sustainable communication networks is to maximize the energy sustainability of the system, such that the harvested energy can fulfill the QoS requirement of users. Specifically, we will

- formulate the network planning and resource management as a joint optimization problem based on statistical historic data; the objective is to deploy the minimal number of green RNs and optimize resource allocation to ensure full network connectivity and users' QoS requirement can be fulfilled with the harvested energy based on the cost threshold; efficient heuristic algorithms are also proposed to offer important guidelines on network deployment and resource management.
- present an analytical model to characterize the dynamic energy charging and discharging processes based on accurate energy harvesting and consumption information; based on developed energy model, a distributed admission control strategy is proposed to achieve high resource utilization while ensuring energy sustainability of the network.

The remaining of this brief is organized as follows. In Chap. 2, we present the background information and literature survey related to sustainable wireless communication networks. We jointly consider relay node placement and sub-carrier allocation issues of sustainable wireless networks in Chap. 3. Chap. 4 introduces an analytical energy model and a distributed admission control strategy based on the model. Finally, We conclude this brief and give a direction of sustainable wireless networks for future research in Chap. 5.

References

1. Y. Zhuang, J. Pan, and L. Cai, "Minimizing energy consumption with probabilistic distance models in wireless sensor networks," in *IEEE INFOCOM*, San Diego, CA, USA, 15-19 Mar. 2010, pp. 1–9.
2. Y. Zhuang, J. Pan, and G. Wu, "Energy-optimal grid-based clustering in wireless microsensor networks," in *IEEE ICDCS Workshop on Wireless Adhoc and Sensor Networking (WWASN)*, 2009.
3. C. Chao and Y. Lee, "A quorum-based energy-saving MAC protocol design for wireless sensor networks," *IEEE Transactions on Vehicular Technology*, vol. 59, no. 2, pp. 813–822, Feb. 2010.
4. G. Miao, N. Himayat, G. Y. Li, and S. Talwar, "Low-complexity energy-efficient scheduling for uplink OFDMA," *IEEE Transactions on Communications*, vol. 60, no. 1, pp. 112–120, Jan. 2012.
5. G. Lim and C. L. J. Jr, "Energy-efficient cooperative beamforming in clustered wireless networks," *IEEE Transactions on Wireless Communications*, vol. 12, no. 3, pp. 1376–1385, Mar. 2013.
6. A. Chamam and S. Pierre, "On the planning of wireless sensor networks: Energy-efficient clustering under the joint routing and coverage constraint," *IEEE Transactions on Mobile Computing*, vol. 8, no. 8, pp. 1077–1086, Aug. 2009.
7. SRI - Center for Uncertainty Quantification in Computer Science & Engineering. [Online]. Available: http://sri-uq.kaust.edu.sa/Pages/greenwireless.aspx
8. Centre for White Space Communications. [Online]. Available: http://www.wirelesswhitespace. org/projects/wind-fi-renewable-energy-basestation.aspx
9. G. P. Fettweis and E. Zimmermann, "ICT energy consumption-trends and challenges," in *WPMC*, Lapland, FI, 8-11 Sep. 2008, pp. 2006–2009.
10. Huawei. [Online]. Available: http://www.greenhuawei.com/green/greenenergy.html
11. Sprint. [Online]. Available: http://www.sprint.com/responsibility/ouroperations/climate_change/renewable-energy.html
12. L. X. Cai, Y. Liu, H. T. Luan, X. Shen, J. W. Mark, and H. V. Poor, "Adaptive resource management in sustainable energy powered wireless mesh networks," in *IEEE Globecom*, Houston, TX, USA, 5-9 Dec. 2011, pp. 1–5.
13. L. X. Cai, H. V. Poor, Y. Liu, T. H. Luan, X. Shen, and J. W. Mark, "Dimensioning network deployment and resource management in green mesh networks," *IEEE Wireless Communications*, vol. 18, no. 5, pp. 58–65, Oct. 2011.
14. Z. Zheng, L. X. Cai, R. Zhang, and X. Shen, "RNP-SA: Joint relay placement and sub-carrier allocation in wireless communication networks with sustainable energy," *IEEE Transactions on Wireless Communications*, vol. 11, no. 10, pp. 3818–3828, Oct. 2012.
15. Z. Zheng, B. Zhang, X. Jia, J. Zhang, and K. Yang, "Minimum AP placement for WLAN with rate adaptation using physical interference model," in *IEEE Globecom*, Miami, FL, USA, 6-10 Dec. 2010, pp. 1–5.
16. J. Zhang, X. Jia, Z. Zheng, and Y. Zhou, "Minimizing cost of placement of multi-radio and multi-power-level access points with rate adaptation in indoor environment," *IEEE Transactions on Wireless Communications*, vol. 10, no. 7, pp. 2186–2195, Jul. 2011.

17. B. Aoun, R. Boutaba, Y. Iraqi, and G. Kenward, "Gateway placement optimization in wireless mesh networks with QoS constraints," *IEEE Journal on Selected Areas in Communications*, vol. 24, no. 11, pp. 2127–2136, Nov. 2006.
18. Y. T. Hou, Y. Shi, H. D. Sherali, and S. F. Midkiff, "On energy provisioning and relay node placement for wireless sensor networks," *IEEE Transactions on Wireless Communications*, vol. 4, no. 5, pp. 2579–2590, Sep. 2005.
19. Z. Cheng, M. Perillo, and W. B. Heinzelman, "General network lifetime and cost models for evaluating sensor network deployment strategies," *IEEE Transactions on Mobile Computing*, vol. 7, no. 4, pp. 484–497, Apr. 2008.
20. F. Wang, D. Wang, and J. Liu, "Traffic-aware relay node deployment: Maximizing lifetime for data collection wireless sensor networks," *IEEE Transactions on Parallel and Distributed Systems*, vol. 22, no. 8, pp. 1415–1423, Aug. 2011.
21. P. G. Brevis, J. Gondzio, Y. Fan, H. V. Poor, J. Thompson, I. Krikidis, and P. J. Chung, "Base station location optimization for minimal energy consumption in wireless networks." in *IEEE VTC*, Budapest, HUN, 15-18 May. 2011, pp. 1–5.
22. S. Saengthong and S. Premrudeepreechacham, "A simple method in sizing related to the reliability supply of small stand-alone photovoltaic systems," in *IEEE PVSC*, Anchorage, AK, USA, 15-22 Sep. 2000, pp. 1630–1633.
23. H. A. M. Maghraby, M. H. Shwehdi, and G. K. Al-Bassam, "Probabilistic assessment of photovoltaic (pv) generation systems," *IEEE Transactions on Power Systems*, vol. 17, no. 1, pp. 205–208, Feb. 2002.
24. E. Lorenzo and L. Navarte, "On the usefulness of stand-alone PV sizing methods," *Progress in Photovoltaics: Research and Applications*, vol. 8, no. 4, pp. 391–409, Aug. 2000.
25. E. Lattanzi, E. Regini, A. Acquaviva, and A. Bogliolo, "Energetic sustainability of routing algorithms for energy-harvesting wireless sensor networks," *Computer Communications*, vol. 30, no. 14-15, pp. 2976–2986, Oct. 2007.
26. T. D. Todd, A. A. Sayegh, M. N. Smadi, and D. Zhao, "The need for access point power saving in solar powered WLAN mesh networks," *IEEE Network*, vol. 22, no. 3, pp. 4–10, May.-Jun. 2008.
27. M. Erol-Kantarci and H. T. Mouftah, "Suresense: sustainable wireless rechargeable sensor networks for the smart grid," *IEEE Wireless Communications*, vol. 19, no. 3, pp. 30–36, Jun. 2012.
28. A. Farbod and T. D. Todd, "Resource allocation and outage control for solar-powered WLAN mesh networks," *IEEE Transactions on Mobile Computing*, vol. 6, no. 8, pp. 960–970, Aug. 2007.
29. L. X. Cai, H. V. Poor, Y. Liu, T. H. Luan, X. Shen, and J. W. Mark, "Dimensioning network deployment and resource management in green mesh networks," *IEEE Wireless Communications*, vol. 18, no. 5, pp. 58–65, Oct. 2011.
30. Z. Zheng, B. Zhang, X. Jia, J. Zhang, and K. Yang, "Minimum AP placement for WLAN with rate adaptation using physical interference model," in *IEEE Globecom*, Miami, FL, USA, 6-10 Dec. 2010, pp. 1–5.
31. P. G. Brevis, J. Gondzio, Y. Fan, H. V. Poor, J. Thompson, I. Krikidis, and P. J. Chung, "Base station location optimization for minimal energy consumption in wireless networks." in *IEEE VTC*, Budapest, HUN, 15-18 May. 2011, pp. 1–5.
32. M. Erol-Kantarci and H. T. Mouftah, "Suresense: sustainable wireless rechargeable sensor networks for the smart grid," *IEEE Wireless Communications*, vol. 19, no. 3, pp. 30–36, Jun. 2012.
33. A. Farbod and T. D. Todd, "Resource allocation and outage control for solar-powered WLAN mesh networks," *IEEE Transactions on Mobile Computing*, vol. 6, no. 8, pp. 960–970, Aug. 2007.
34. E. Lattanzi, E. Regini, A. Acquaviva, and A. Bogliolo, "Energetic sustainability of routing algorithms for energy-harvesting wireless sensor networks," *Computer Communications*, vol. 30, no. 14-15, pp. 2976–2986, Oct. 2007.

Chapter 2
Background and Literature Survey

In the last decade, a large number of works on energy-efficient, high-quality and low-cost wireless access services have shown up [1–3]. Generally, these solutions can be divided into two classes, namely, customer-oriented and infrastructure-oriented solutions. Customer devices, e.g., wireless sensor nodes [4] and mobile terminals, usually are powered by batteries. Thus, the research objective of customer-oriented solutions mainly focus on improving energy efficiency to prolong the battery lifetime by various methods including energy-efficient software applications [5], hardware design [6], and protocol improvements [7]. Compared with customer devices, the network infrastructure contributes to the dominant portion of the total energy consumption of the system. For example, the BSs consume 60–80 % of the network's energy consumption [8, 9]. Therefore, it is more important to reduce energy consumption of the infrastructure in order to improve the energy efficiency of the overall system. To sustain the wireless operations, one promising solution is to use green energy to power the infrastructure network devices. In such a sustainable network, the research objective and performance metric are changed from energy efficiency to energy sustainability, i.e., to ensure harvested energy can sustain the normal network operations. We categorize the existing works in the literature related on sustainable wireless networks into three research issues: (1) network planning, (2) energy modeling, and (3) resource allocation.

2.1 Network Planning

Network planning has been extensively studied in the context of different wireless networks, including cellular networks, IEEE 802.16 WiMAX, and sensor networks [10–12]. Network planning is usually formulated as device deployment optimization problems, aiming at maximizing the network capacity [13, 14] or minimizing the cost of device deployment and/or network operation [12, 15, 16]. According to the methodologies to solve the optimization problem, these works can be further classified into two types, i.e., continuous and discrete cases. In the continuous case, it is assumed that there is no physical constraints and wireless network devices can be deployed at any location of the network region [17, 18]. Such problems can be solved by using some optimization algorithms like direct search and quasi-Newton methods [19]. However, in reality, wireless devices usually can only be placed at some candidate locations due to the physical constraints. Such problems can be formulated as the discrete cases of device deployment problems. The discrete problems are normally modeled as a mixed integer optimization problem to find out the optimal placement of devices in a given region (or among a set of users), such that all the users in the region can be served by the deployed network devices [15, 20, 21]. In [21], a relay node placement problem is investigated with the physical constraints of sensor nodes. In [15], how to place the minimal number of APs is studied under the physical and protocol interference models; and it is found that the underlying interference models have a significant effect on the AP placement problem. In [16], the optimization of base stations' number and locations is investigated in order to minimize the energy consumption of a cellular network, considering a practical case of non-uniform user distributions.

There have been limited works on network planning in sustainable wireless networks, which mainly focus on how to minimize the cost and network outage, i.e., some green wireless devices do not have sufficient energy to support normal operation or data transmission. The possibility and advantages of deploying a sustainable energy powered wireless system are reported in [22]. It is shown that solar or wind powered APs provide a cost-effective solution in wireless local area networks (WLANs), especially for APs installed in off-grid locations. In [1, 23], the traditional AP placement problem is revisited with sustainable power supplies. Their work focuses on placing a minimal number of green energy powered APs on a set of candidate locations to ensure that the harvested energy is sustainable to serve wireless users and fulfill their QoS requirement. The minimum-cost placement of solar-powered data collection BSs is considered in [24]. BSs are placed in a wireless sensor network, such that the outage-free operation of the sensor nodes can be obtained. In [25], authors jointly consider allocating transmitting power and deploying the green APs based on the harvested energy. In this work, a closed form power allocation scheme and an AP placement metric are proposed, and their theoretical analysis shows a dramatically improvement on overall throughput by using the proposed scheme.

2.2 Energy Modeling

One of the effective methods to prolong the battery life is to enhance the energy efficiency by designing an accurate analytic energy model [26–28]. In [29], a model which integrates typical WSNs transmission and reception modules with realistic battery models is proposed. Based on the battery models, they propose two battery power-conserving schemes for two M-ary orthogonal modulations. In [30], authors focus on designing time division multiple access medium access control protocols for healthcare applications in wireless body-area monitoring networks. They find that the proposed schemes can extend the lifetime of sensor nodes for the wireless body-area monitoring networks based on the theoretical and simulations.

The first addressed issue in many works related to green energy is sustainable wireless sensor networks with renewable energy [31–33]. In [34] and [35], authors show that such kind of prototypes can achieve near-perpetual operation of a sensor node. In WLAN mesh networks, the solar/wind powered AP is believed to be a more efficient method to save energy than energy efficient schemes in traditional AP, especially when the traditional power supply is not available. Different from traditional energy resources, we need to consider the inherently dynamic characteristics in both energy charging and discharging processes. Therefore, it is essential to characterize the variations in the analytical model of energy conditions. In [36], authors design a framework to model the remaining power of sensor nodes with and without green energy, and then the expression of network lifetime can be derived based on the energy model. In [37], the transmission policies for rechargeable nodes are considered to maximize the short term throughput, which refers to the amount of data transmitted in a finite time horizon. Based on the renewable energy model with discrete packets of energy arrivals, their proposed algorithm can successfully generate the optimal transmission policy, which can achieve the maximum short-term throughput and the minimum transmission completion time. In [38], the sustainable wireless rechargeable sensor network is proposed with mobile chargers charging multiple sensors from candidate locations. After that, an optimization model is developed to minimize the selected number of locations based on the energy recharging requirement of the sensors. Other works, such as [39, 40], mainly focus on the battery capacity and solar panel size of the BSs or APs, with an objective to mitigate the network outage by using the minimal cost of energy according to the recorded historical solar insolation traces.

2.3   Resource Allocation

Resource allocation is one of the most crucial methods to enhance the resource utilization of wireless networks [41–44]. Many works have been studied in various aspects of resource allocation, which include traffic scheduling and routing [45–47], optimal power management [48–50], energy efficient communication and cooperation [51–53], and adaptive sleep control of mobile devices [54–56], etc. Resource allocation [12, 57, 58] in infrastructure network can be formulated as an optimization problem such that the network performance, e.g. maximizing network throughput and maximizing network lifetime, etc., with fixed yet limited energy resource in traditional wireless networks is maximized, under various constraints including network connectivity, throughput, energy consumption and etc. The energy in these works is normally considered as a limited resource, thus these works generally target at maximizing the energy efficiency.

In sustainable wireless networks, the energy is sustainable in the long term yet dynamic in the short term, which may lead to intermittent energy supply in wireless network infrastructure devices [1]. Moreover, since the green wireless devices highly depend on their locations, which leads to uneven distribution of charging capabilities.

Thus, in order to balance the harvested energy and traffic demand, we should concern these characteristics and challenges of sustainable wireless networks.

So far, only a few works on resource management in wireless networks with green energy focus on maximizing the network sustainability, and most existing works aim at mitigating the node outage or minimizing the cost. In [39], the work focuses on solar panel sizing problem of the BS or APs based on the historical solar insolation traces, such that the network outage can be mitigated and the cost can be minimized. In [59], the problem of traffic scheduling for infrastructure of vehicular wireless networks is formulated into a mixed integer linear program with minimizing energy consumption as objective. In [60], authors propose a framework by jointly considering integrated admission control and routing under the multi-hop radio networks powered by green energy. Then, routing algorithms are proposed to improve network performance by using available energy. In [61], statistical power saving mechanism is proposed under solar-powered WLAN mesh networks. To balance the energy consumption with energy charging capability for each node, a control algorithm is developed to match the future load conditions and solar insolation for maintaining outage-free operations of the node.

References

1. L. X. Cai, H. V. Poor, Y. Liu, T. H. Luan, X. Shen, and J. W. Mark, "Dimensioning network deployment and resource management in green mesh networks," *IEEE Wireless Communications*, vol. 18, no. 5, pp. 58–65, Oct. 2011.
2. Y. Chen, S. Zhang, S. Xu, and G. Y. Li, "Fundamental trade-offs on green wireless networks," *IEEE Communications Magazine*, vol. 49, no. 6, pp. 30–37, Jun. 2011.
3. M. Asefi, J. W. Mark, and X. Shen, "A mobility-aware and quality-driven retransmission limit adaptation scheme for video streaming over VANETs," *IEEE Transactions on Wireless Communications*, vol. 11, no. 5, pp. 1817–1827, May. 2012.
4. I. F. Akyildiz, W. Su, Y. Sankarasubramaniam, and E. Cayirci, "Wireless sensor networks: a survey," *Computer Networks*, vol. 38, no. 4, pp. 393–422, Mar. 2002.
5. N. A. Pantazis and D. D. Vergados, "A survey on power control issues in wireless sensor networks," *IEEE Communications Surveys & Tutorials*, vol. 9, no. 4, pp. 86–107, 2007.
6. M. Hempstead, M. J. Lyons, D. Brooks, and G. Y. Wei, "Survey of hardware systems for wireless sensor networks," *Journal of Low Power Electronics*, vol. 4, no. 1, pp. 11–20, Apr. 2008.
7. K. Akkaya and M. Younis, "A survey on routing protocols for wireless sensor networks," *Ad hoc networks*, vol. 3, no. 3, pp. 325–349, May. 2005.
8. M. A. Marsan, L. Chiaraviglio, D. Ciullo, and M. Meo, "Optimal energy savings in cellular access networks," in *IEEE ICC Workshops*, Dresden, DE, 14–18 Jun. 2009.
9. E. Oh, B. Krishnamachari, X. Liu, and Z. Niu, "Toward dynamic energy-efficient operation of cellular network infrastructure," *IEEE Communications Magazine*, vol. 49, pp. 56–61, Jun. 2011.
10. J. Pan, L. Cai, Y. Shi, and X. Shen, "Optimal base-station locations in two-tiered wireless sensor networks," *IEEE Transactions on Mobile Computing*, vol. 4, no. 5, pp. 458–473, Sep.-Oct. 2005.
11. M. Soleimanipour, W. Zhuang, and G. H. Freeman, "Optimal resource management in wireless multimedia wideband CDMA systems," *IEEE Transactions on Mobile Computing*, vol. 1, no. 2, pp. 143–160, Apr.-Jun. 2002.

12. B. Lin, P. Ho, L. Xie, X. Shen, and J. Tapolcai, "Optimal relay station placement in broadband wireless access networks," *IEEE Transactions on Mobile Computing*, vol. 9, no. 2, pp. 259–269, Feb. 2010.

13. X. Ling and K. L. Yeung, "Joint access point placement and channel assignment for 802.11 wireless LANs," *IEEE Transactions on Wireless Communications*, vol. 5, no. 10, pp. 2705–2711, Oct. 2006.

14. M. Unbehaun and M. Kamenetsky, "On the deployment of picocellular wireless infrastructure," *IEEE Wireless Communications*, vol. 10, no. 6, pp. 70–80, Dec. 2003.

15. Z. Zheng, B. Zhang, X. Jia, J. Zhang, and K. Yang, "Minimum AP placement for WLAN with rate adaptation using physical interference model," in *IEEE Globecom*, Miami, FL, USA, 6-10 Dec. 2010, pp. 1–5.

16. P. G. Brevis, J. Gondzio, Y. Fan, H. V. Poor, J. Thompson, I. Krikidis, and P. J. Chung, "Base station location optimization for minimal energy consumption in wireless networks." in *IEEE VTC*, Budapest, HUN, 15-18 May. 2011, pp. 1–5.

17. W. Zhang, G. Xue, and S. Misra, "Fault-tolerant relay node placement in wireless sensor networks: Problems and algorithms," in *IEEE INFOCOM*, Anchorage, AK, USA, 6-12 May. 2007, pp. 1649–1657.

18. X. Han, X. Cao, E. L. Lloyd, and C. C. Shen, "Fault-tolerant relay node placement in heterogeneous wireless sensor networks," *IEEE Transactions on Mobile Computing*, vol. 9, no. 5, pp. 643–656, May. 2010.

19. Z. Wei, G. Li, and L. Qi, "New quasi-newton methods for unconstrained optimization problems," *Applied Mathematics and Computation*, vol. 175, no. 2, pp. 1156–1188, Apr. 2006.

20. I. K. Fu, W. H. Sheen, and F. C. Ren, "Deployment and radio resource reuse in IEEE 802.16j multi-hop relay network in manhattan-like environment," in *IEEE ICICS*, Meritus Mandarin Hotel, Singapore, 10-13 Dec. 2007, pp. 1–5.

21. S. Misra, S. D. Hong, G. Xue, and J. Tang, "Constrained relay node placement in wireless sensor networks: Formulation and approximations," *IEEE/ACM Transactions on Networking*, vol. 18, no. 2, pp. 434–447, Apr. 2010.

22. A. Sayegh, T. D. Todd, and M. Smadi, "Resource allocation and cost in hybrid solar/wind powered WLAN mesh nodes," *Wireless Mesh Networks: Architectures and Protocols*, pp. 167–189, 2007.

23. Z. Zheng, L. X. Cai, M. Dong, X. Shen, and H. V. Poor, "Constrained energy-aware ap placement with rate adaptation in WLAN mesh networks," in *IEEE GLOBECOM*, Houston, TX, USA, 5-9 Dec. 2011, pp. 1–5.

24. S. A. Shariatmadari, A. A. Sayegh, and T. D. Todd, "Energy aware basestation placement in solar powered sensor networks," in *IEEE WCNC*, Sydney, AUS, 18-21 Apr. 2010, pp. 1–6.

25. X. Zhang, Z. Zheng, J. Liu, X. Shen, and L. Xie, "Optimal power allocation and AP deployment in green wireless cooperative communications," in *IEEE GLOBECOM*, Anaheim, CA, USA, 3-7 Dec. 2012, pp. 4000–4005.

26. B. Kan, L. Cai, H. Zhu, and Y. Xu, "Accurate energy model for WSN node and its optimal design," *Journal of Systems Engineering and Electronics*, vol. 19, no. 3, pp. 427–433, Jun. 2008.

27. C. Ma and Y. Yang, "A battery-aware scheme for routing in wireless ad hoc networks," *IEEE Transactions on Vehicular Technology*, vol. 60, no. 8, pp. 3919–3932, Oct. 2011.

28. J. Vazifehdan, R. V. Prasad, M. Jacobsson, and I. Niemegeers, "An analytical energy consumption model for packet transfer over wireless links," *IEEE Communications Letters*, vol. 16, no. 1, pp. 30–33, Jan. 2012.

29. Q. Tang, L. Yang, G. B. Giannakis, and T. Qin, "Battery power efficiency of PPM and FSK in wireless sensor networks," *IEEE Transactions on Wireless Communications*, vol. 6, no. 4, pp. 1308–1319, Apr. 2007.

30. H. Su and X. Zhang, "Battery-dynamics driven TDMA MAC protocols for wireless body-area monitoring networks in healthcare applications," *IEEE Journal on Selected Areas in Communications*, vol. 27, no. 4, pp. 424–434, May. 2009.

31. T. J. Kazmierski, G. V. Merrett, L. Wang, B. M. Al-Hashimi, A. S. Weddell, and I. N. Ayala-Garcia, "Modeling of wireless sensor nodes powered by tunable energy harvesters: HDL-based approach," *IEEE Sensors Journal*, vol. 12, no. 8, pp. 2680–2689, Jun. 2012.

32. A. S. Weddell, G. V. Merrett, T. J. Kazmierski, and B. M. Al-Hashimi, "Accurate supercapacitor modeling for energy harvesting wireless sensor nodes," *IEEE Transactions on Circuits and Systems II: Express Briefs*, vol. 58, no. 12, pp. 911–915, Dec. 2011.

33. P. T. Venkata, S. N. A. U. Nambi, R. V. Prasad, and I. Niemegeers, "Bond graph modeling for energy-harvesting wireless sensor networks," *Computer*, vol. 45, no. 9, pp. 31–38, Sep. 2012.

34. J. Taneja, J. Jeong, and D. Culler, "Design, modeling, and capacity planning for micro-solar power sensor networks," in *IPSN SPOTS*, Apr. 2008.

35. V. Raghunathan, A. Kansal, J. Hsu, J. Friedman, and M. Srivastava, "Design considerations for solar energy harvesting wireless embedded systems," in *IPSN*. Piscataway, NJ, USA: IEEE Press, 2005, p. 64.

36. K. Ramachandran and B. Sikdar, "A population based approach to model the lifetime and energy distribution in battery constrained wireless sensor networks," *IEEE Journal on Selected Areas in Communications*, vol. 28, no. 4, pp. 576–586, Apr. 2010.

37. K. Tutuncuoglu and A. Yener, "Optimum transmission policies for battery limited energy harvesting nodes," *IEEE Transactions on Wireless Communications*, vol. 11, no. 3, pp. 1180–1189, Mar. 2012.

38. M. Erol-Kantarci and H. T. Mouftah, "Suresense: sustainable wireless rechargeable sensor networks for the smart grid," *IEEE Wireless Communications*, vol. 19, no. 3, pp. 30–36, Jun. 2012.

39. G. H. Badawy, A. A. Sayegh, and T. D. Todd, "Energy provisioning in solar-powered wireless mesh networks," *IEEE Transactions on Vehicular Technology*, vol. 59, no. 8, pp. 3859–3871, Oct. 2010.

40. M. S. Zefreh, G. H. Badawy, and T. D. Todd, "Position aware node provisioning for solar powered wireless mesh networks," in *IEEE GLOBECOM*, Miami, FL, USA, 6-10 Dec. 2010, pp. 1–6.

41. W. Tuttlebee, S. Fleccher, D. Lister, T. Farrell, and J. Thompson, "Saving the plannet – the rationale, realities and research of green radio," *International Transfer Pricing Journal*, vol. 4, no. 3, Sep. 2010.

42. C. Han, T. Harrold, S. Armour, I. Krikidis, S. Videv, P. M. Grant, H. Haas, J. S. Thompson, I. Ku, C. X. Wang, T. A. Le, M. R. Nakhai, J. Zhang, and L. Hanzo, "Green radio: radio techniques to enable energy-efficient wireless networks," *IEEE Communications Magazine*, vol. 49, no. 6, pp. 46–54, Jun. 2011.

43. P. Grant and S. Fletcher, "Mobile basestations: reducing energy," *Engineering & Technology Magazine*, vol. 6, no. 2, Feb. 2011.

44. H. Zhang, A. Gladisch, M. Pickavet, Z. Tao, and W. Mohr, "Energy efficiency in communications," *IEEE Communications Magazine*, vol. 48, no. 11, pp. 48–49, Nov. 2010.

45. L. X. Cai, L. Cai, X. Shen, and J. W. Mark, "Resource management and QoS provisioning for IPTV over mmwave-based WPANs with directional antenna," *Mob. Netw. Appl.*, vol. 14, no. 2, pp. 210–219, 2009.

46. A. Liu, Z. Zheng, C. Zhang, Z. Chen, and X. Shen, "Secure and energy-efficient disjoint multi-path routing for WSNs," *IEEE Trans. on Vehicular Technology*, vol. 61, no. 7, pp. 3255–3265, 2012.

47. Y. Liu, L. X. Cai, and X. Shen, "Spectrum-aware opportunistic routing in multi-hop cognitive radio networks," *IEEE J. Selected Areas of Communications*, vol. 30, no. 10, pp. 1958–1969, Nov. 2012.

48. H. T. Cheng and W. Zhuang, "QoS-driven MAC-layer resource allocation for wireless mesh networks with non-altruistic node cooperation and service differentiation," *IEEE Transactions on Wireless Communications*, vol. 8, no. 12, pp. 6089–6103, Dec. 2009.

49. R. L. Cruz and A. V. Santhanam, "Optimal routing, link scheduling and power control in multi-hop wireless networks," in *IEEE INFOCOM*, vol. 1, Mar. 2003, pp. 702–711.

50. M. Veluppillai, J. W. Mark, and X. Shen, "Performance analysis and power allocation for M-QAM cooperative diversity systems," *IEEE Transactions on Wireless Communications*, vol. 9, no. 3, pp. 1237–1247, Mar. 2010.

51. X. Zhang, L. Xie, and X. Shen, "Energy-efficient transmission and bit allocation schemes in wireless sensor networks," *Int. J. of Sensor Networks (IJSNET)*, vol. 11, no. 4, pp. 241–249, 2012.

52. B. Cao, Q. Zhang, J. Mark, L. Cai, and H. Poor, "Toward efficient radio spectrum utilization: user cooperation in cognitive radio networking," *IEEE Network*, vol. 26, no. 4, pp. 46–52, Jul. 2012.

53. M. S. Alam, J. W. Mark, and X. Shen, "Relay selection and resource allocation for multi-user cooperative OFDMA networks," *IEEE Transactions on Wireless Communications*, vol. 12, no. 5, pp. 2193–2204, May. 2013.

54. B. J. Choi and X. S. Shen, "Adaptive asynchronous sleep scheduling protocols for delay tolerant networks," *IEEE Transactions on Mobile Computing*, vol. 10, no. 9, pp. 1283–1296, Sep. 2011.

55. Y. Dong and D. Yau, "Adaptive sleep scheduling for energy-efficient movement-predicted wireless communication," in *IEEE International Conference on Network Protocols*, Boston, MA, USA, 6-9 Nov. 2005, pp. 1–10.

56. J. H. Jeon, H. J. Byun, and J. T. Lim, "Joint contention and sleep control for lifetime maximization in wireless sensor networks," *IEEE Communications Letters*, vol. 17, no. 2, pp. 269–272, Mar. 2013.

57. Y. Shi, Y. T. Hou, and A. Efrat, "Algorithm design for base station placement problems in sensor networks," in *ACM QSHINE*, Waterloo, ON, CA, 7-9 Aug. 2006.

58. H. Liang, L. X. Cai, D. Huang, X. Shen, and D. Peng, "A SMDP-based service model for inter-domain resource allocation in mobile cloud networks," *IEEE Transactions on Vehicular Technology*, vol. 61, no. 5, pp. 2222–2232, Jun. 2012.

59. A. A. Hammad, G. H. Badawy, T. D. Todd, A. A. Sayegh, and D. Zhao, "Traffic scheduling for energy sustainable vehicular infrastructure," in *IEEE GLOBECOM*, Miami, FL, USA, 6-10 Dec. 2010, pp. 1–6.

60. L. Lin, N. B. Shroff, and R. Srikant, "Asymptotically optimal energy-aware routing for multihop wireless networks with renewable energy sources," *IEEE/ACM Transactions on Networking*, vol. 15, no. 5, pp. 1021–1034, Oct. 2007.

61. A. Farbod and T. D. Todd, "Resource allocation and outage control for solar-powered WLAN mesh networks," *IEEE Transactions on Mobile Computing*, vol. 6, no. 8, pp. 960–970, Aug. 2007.

Chapter 3
Joint Relay Placement and Sub-carrier Allocation in Sustainable Wireless Networks

Nowadays, the increasing concern on environmental protection has boosted the research of sustainable wireless networks construction. It is anticipated that wireless devices including BSs and RNs will be powered by green energy to sustain the operations of wireless networks. However, green energy is normally sustainable yet variable, which makes it challenging to exploit green energy sources in the deployment and management of a sustainable wireless network. In this chapter, the network planning and resource management issues are re-visited in the context of sustainable wireless networks based on statistical historic data. The objective is to deploy the minimal number of green RNs in the sustainable wireless networks, such that the harvested energy can fulfill the throughput requirement of users based on the cost threshold. To this end, we jointly consider the RN placement and sub-carrier allocation problem, and formulate it as a mixed integer non-linear programming (MINLP) problem. As the formulated problem is NP-complete, we propose two low-complexity heuristic algorithms, namely RNP-SA with top-down/bottom-up algorithms (RNP-SA-t/b) to solve it. The extensive simulation results show that our proposed algorithms can efficiently solve the problem with low time complexity, and offer important guidelines on network deployment and resource management in a sustainable wireless network.

The remainder of the chapter is organized as follows. Introduction is presented in Sect. 3.1. The system model of a two-tiered wireless network with sustainable energy is described in Sect. 3.2. Based on the system model, a joint RN placement and sub-carrier allocation problem is formulated as an MINLP problem in Sect. 3.3. Two heuristic algorithms are proposed in Sect. 3.4. The performance of our algorithms is compared with that of a greedy algorithm in different network scenarios in Sect. 3.5, followed by conclusions in Sect. 3.6.

3.1 Introduction

As promising research issues, there have been some works addressed the network planning and resource allocation issues in sustainable wireless networks. However, previous works try to improve the network sustainability by network planning and resource allocation separately, which ignore mutual impact on each other to achieve long term energy sustainability. In this chapter, we focus on jointly considering the network planning and resource allocation problems in sustainable wireless networks, with the bandwidth, cost and energy sustainability constraints. Specifically, the relay node placement and sub-carrier allocation problem is studied in the context of sustainable wireless networks, where the green BSs and RNs can harvest energy from the natural resources. We consider a realistic two-tiered network scenario, where green BSs are already deployed and there are a set of candidate locations for placing green RNs. Our objective is to minimize the number of placed green RNs with allocating appropriate number of sub-carriers to each green BS and RN, meanwhile ensuring that the network can be connected and the traffic demand of users can be fulfilled by harvested energy based on the cost threshold. Then, we formulate the joint RNP-SA problem as an MINLP problem, which is NP-hard. To provide simple yet efficient solutions, we propose two low-complexity and effective heuristic algorithms. A performance metric, called Sub-carrier and Traffic over Rate (STR), is introduced to characterize the throughput and energy demand of links. Based on STR, the proposed algorithms iteratively select candidate locations to place RNs and connect users to the deployed RNs until any of the cost threshold, energy sustainability constraint or the bandwidth requirement of users cannot be satisfied. Finally, we obtain the minimal number of RNs to fulfill the users' traffic demand by letting each RN to serve as many users as possible.

3.2   System Model

In this chapter, a two-tiered sustainable wireless network consisting of BSs, RNs, and wireless users is considered as shown in Fig. 4.1. We consider the network scenario with explosive growth of network density and traffic intensity, thus it requires the deployment of extra RNs in the existing cellular network to improve the QoS provisioning and network capacities. In the existing cellular network, a set of BSs have been installed and a number of candidate locations are given for RNs placement. BSs and RNs are powered by green energy, e.g., solar panels or wind turbines, which can harvest energy from natural resources. Due to the hardware and cost constraints, the personal devices of wireless users are powered by batteries. Different from traditional energy sources, green energy sources are normally dynamic and highly dependent on its position, local weather and time, thus green BSs and RNs distributed at different locations have various charging capabilities. This kind of network scenario is common in real life. For instance, a community with a set of green BSs, where each building with people working or living in it refers to a user and RNs can be installed on some candidate locations, which may be the roof of certain buildings. The wind speed level or solar insolation of candidate locations or BSs and the traffic demand of users can be obtained from historical statistics data. A summary of the mathematical notations used in this chapter is given in Table 3.1.

Table 3.1 The notations for Chap. 3	

V	The total set of nodes, i.e., BSs, RNs, and users
E	The set of communication links between any two nodes
U	The set of users
R	The set of RNs
B	The set of BSs
S	The set of sub-carriers
L	The set of candidate locations
D_{xy}	The distance between a pair of nodes, x and y
P_x^T	The transmission power of node x
SNR_{xy}	The received signal to noise ratio of link (x, y)
α	The path loss exponent
N	The background noise
c_{xy}	The achievable data rate of link (x, y)
W	The sub-carrier bandwidth
S_{xy}	The set of sub-carriers allocated to link (x, y)
$|S_{xy}|$	The number of allocated sub-carriers
SNR_{yx}	Threshold of received signal strength
I_x	The interference collision set of node x
β_{xy}	Achieved flow throughput from node x to y
β'_{xy}	User's throughput requirement
$\mathcal{E}_x^-$	Consumed energy of node x during a unit time
$\mathcal{E}_x^+$	Harvested energy of node x during a unit time
$|R|$	The cost of deployed RNs
$|R'|$	The determined cost threshold
e_{xy}	Connection status between node x and y
t_{xy}	The active time of link (x, y) during a unit time
γ_x^{up}	The uplink throughput of node x
γ_x^{dn}	The downlink throughput of node x
P_x^+	The average energy charging rate of node x
P_x^R	The power consumption of user x to receive and decode the signal
$T_I(x)$	The active time of the allocated sub-carriers of user x
γ_u	The summation of γ_u^{up} and γ_u^{dn}
(u, r, b)	(u, r) and (r, b)
STR_{urb}	$STR_{ur} + STR_{rb}$
b_x	The BS that x is connected including the case that x is b_x for $x \in B$
γ_z^I	Summation of users' traffic demand within z's interference range, $z \in L$

In the network, each RN connects with a BS, and each user can be associated with either a BS or a RN. Each RN provides service to its users and helps to relay users' traffic to a BS; while the BS transmits the traffic to the user's destination cell through the wired backbone network. Thus, if a user connects to a RN, a two-hop transmission is involved; and a direct link transmission is used if a user is associated with a BS. There are total S sub-carriers in the network. If and only if the sub-carriers by users are sufficiently far away from each other, users can reuse the sub-carrier in the space domain, and the interference is negligible. Since users distributed all over the network have various traffic demand and green devices over the network have

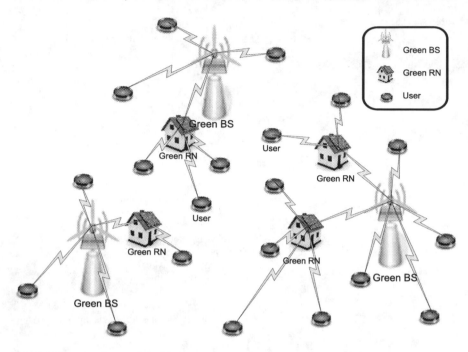

Fig. 3.1 Two-tiered network architecture

different charging capacities, thus it is essential to appropriately select positions to deploy green RNs Fig. 3.1.

We use a network communication graph $G(V = U \cup R \cup B; E)$ to represent the network topology. We can obtain the received signal to noise ratio of link (x, y), denoted SNR_{xy}, as follows,

$$SNR_{xy} = \frac{P_x^T \cdot D_{xy}^{-\alpha}}{N}. \tag{3.1}$$

Then, we have

$$c_{xy} = |S_{xy}| W \log_2 (1 + SNR_{xy}). \tag{3.2}$$

The set of sub-carriers allocated for each link can be expressed as

$$\left\{ \begin{array}{l} S_{xy} = \begin{pmatrix} s_1 \\ s_2 \\ \vdots \\ s_m \end{pmatrix}, \forall (x, y) \in E \\ m = |S_{total}| \\ s_i \in \{0, 1\}, \end{array} \right. \tag{3.3}$$

where $s_i = 1$ means that sub-carrier s_i is allocated to link (x, y), and $s_i = 0$ otherwise.

In the network, users associated with the BS/RN are allocated the same set of sub-carriers as their BS/RN for data transmission. If the received signal strength of x from node y, denoted SNR_{yx}, is over a threshold θ, the node x is located in the interference range of y,

$$SNR_{yx} \geq \theta. \tag{3.4}$$

The interference collision set of y, called I_y, is defined as the set of nodes located within the interference range of y. There is no concurrent transmission at the same time allowed in the same interference collision set.

3.3 Problem Formulation

In this chapter, we aim at deploying the minimal number of RNs into the network and allocating an appropriate number of sub-carriers to each BS/RN to provide full coverage, i.e., any user in the network is served by either a BS or a RN, and users' throughput requirement can be fulfilled by the harvested energy along with the allocated sub-carriers based on the cost threshold. Normally, we deploy more RNs to the network when traffic demand of users can not be sustained by the charging capabilities of BSs. With the help of RNs, BSs can reduce the energy consumption to sustain the network and users can achieve a higher transmission rate. By allocating more sub-carriers, the link will consume more energy to reach a higher throughput with a wider spectrum band. The limited charging capabilities of green BSs/RNs make essential to jointly consider users' traffic demand and the charging capabilities of the BSs/RNs in sub-carrier allocation. For example, if a BS/RN has a low energy level, the energy will be quickly used up and the BS/RN will be out of service by allocating a larger number of sub-carriers. Thus, we need to carefully formulate a joint optimization problem to achieve satisfactory sustainability performance.

Given a set of users (U), a set of existing BSs (B), the available set of sub-carriers, the traffic demand of each user, the charging capability of each BS, the expected charging capabilities of RNs at various candidate locations and the cost threshold, we can formulate the RNP-SA as an MINLP problem,

$$
\begin{aligned}
&\textit{Minimize} \quad |R| \\
&\textit{Subject to:} \quad \sum_{x \in R \cup B} e_{ux} = 1, \quad \forall u \in U \\
&\qquad\qquad\quad \sum_{b \in B} e_{rb} = 1, \qquad \forall r \in R \\
&\qquad\qquad\quad \beta_{ux} \geq \beta'_{ux}, \qquad \forall e_{ux} = 1 \\
&\qquad\qquad\quad \mathcal{E}_x^+ \geq \mathcal{E}_x^-, \qquad\quad \forall x \in R \cup B \\
&\qquad\qquad\quad |R| \leq |R'| \\
&\qquad\qquad\quad e_{xy} \in \{0, 1\}, \qquad \forall x, y \in V
\end{aligned}
\tag{3.5}
$$

In the first and second constraints, each user is only allowed to be associated with one BS or RN, and each RN can only connect to one BS. In the third constraint, the achieved throughput of link (x, y), β_{xy}, should be sufficient to fulfill users' traffic demand, β'_{xy}. In fourth constraint, the harvested energy, $\mathcal{E}_x^+$, should be enough to fulfill the energy consumption of users' traffic demand, $\mathcal{E}_x^-$. In the fifth constraint, the cost of deployed RNs should not exceed the cost threshold. Finally, we define $e_{xy} = 1$ to represent x connects with y, and $e_{xy} = 0$ otherwise.

3.3.1 QoS and Energy Sustainability Constraints

In this subsection, we specify the energy and throughput constraints in Eq. 3.5. In the two-hop uplink transmissions including users, RNs and BSs, the input and output traffic should be the same, which can be expressed as

$$\sum_{u \in U} (e_{ur} c_{ur} t_{ur}) = \sum_{b \in B} e_{rb} c_{rb} t_{rb}, \forall r \in R. \tag{3.6}$$

In order to fulfill the uplink traffic demand of users, we need to ensure

$$\sum_{r \in R} e_{ur} c_{ur} t_{ur} + \sum_{b \in B} e_{ub} c_{ub} t_{ub} \geq \gamma_u^{up}, \forall u \in U, \tag{3.7}$$

where the user is associated with a RN for $e_{ur} = 1$ and $e_{ub} = 0$, and the user is connected to a BS for $e_{ub} = 1$ and $e_{ur} = 0$. We use the first item in the left hand side (LHS) of (3.7) to express the achieved uplink throughput of user u if the user is connected to a RN; and the second item in the LHS to indicate the achieved throughput if the user is connected to a BS.

Since the input and output traffic of RNs should be the same, we can obtain

$$\sum_{u \in U} (e_{ru} c_{ru} t_{ru}) = \sum_{b \in B} e_{br} c_{br} t_{br}, \forall r \in R. \tag{3.8}$$

In order to fulfill the traffic demand of users, we have to guarantee

$$\sum_{r \in R} e_{ru} c_{ru} t_{ru} + \sum_{b \in B} e_{bu} c_{bu} t_{bu} \geq \gamma_u^{dn}, \forall u \in U. \tag{3.9}$$

We have to ensure that the harvested energy should be enough for the energy required to send and receive the traffic to and from other nodes, such that the energy sustainability can be ensured.

$$P_x^+ \geq P_x^T \sum_{(x,y) \in E} t_{xy} + P_x^R \sum_{(y,x) \in E} t_{yx}, \forall x \in R \cup B. \tag{3.10}$$

Finally, the concurrent transmissions are not allocation within the same interference collision set to avoid harmful interference, which can be expressed as

$$T_I(x) = \sum_{y \in I_x} \sum_{z \in V} (t_{yz} S_{yz} + t_{zy} S_{zy}) \leq 1, \forall x \in V. \tag{3.11}$$

The equation shows that the total active time of links in the same collision set should not exceed a unit time.

3.4 RNP-SA Algorithms

As the subproblems of RNP-SA problem, e.g., the RNP problem, are well-know NP-hard problems, the RNP-SA problem is NP-hard and normally has no efficient polynomial time solution to address the problem in [1]. Thus, we try to design efficient heuristic algorithms with low time complexity for the RNP-SA problem in this chapter. As such, we investigate the performance metric which has significant impact on throughput and energy constraints. If a RN is placed close to a BS, the energy consumption of the BS can be significantly decreased as the BS only needs to communicate with the RN, but the throughput gain is limited. If a RN is deployed in heavy traffic load area and far away from the BS, much more traffic requirement can be fulfilled as users can communicate with the RN, but the decrease of energy consumption is little. Therefore, it is critical to balance the throughput gain and energy consumption of BSs by carefully determining the location to deploy green RNs into the network. We design a performance metric, called the Sub-carrier and Traffic over Rate (STR), to determine whether and where a RN is needed to be deployed by jointly considering the energy and bandwidth requirement. Two low-complexity heuristic algorithms, RNP-SA with top-down/bottom-up algorithms (RNP-SA-t/b), are proposed according to the STR metric.

We introduce the overview of our algorithms in Sect. 3.4.1. Then, the designed performance metric is presented in Sect. 3.4.2. We describe the detail of RNP-SA-t/b algorithms in Sects. 3.4.3 and 3.4.4. At last, we analyze the time complexity of proposed RNP-SA-t/b algorithms in Sect. 3.4.5.

3.4.1 Algorithms Overview

Our target is placing the minimal number of RNs to guarantee that the throughput requirement of users can be fulfilled under the constraints of energy sustainability and cost. One of the intuitive methods is to place the RNs into the candidate locations with the heaviest traffic load to meet users' throughput requirement. However, when the candidate locations with the heaviest traffic load are far away from BSs, BSs still need to consume a lot of energy to communicate with RNs. Another intuitive method is to place RNs close to the BSs to relieve the energy consumption of BSs.

Nonetheless, wireless users communicating with RNs have the similar throughput as with BS, thus the traffic demand of users may not be fulfilled under this strategy. Therefore, we design performance metric to characterize the throughput and energy demand of each user associated with a RN or BS, such that the energy consumption and throughput can be balanced for RNs deployment.

Based on the performance metric, two low-complexity heuristic algorithms are proposed to minimize the number of deployed RNs with top-down and bottom-up methods. In the top-down algorithm, RNs are deployed in all candidate locations at first. We connect all links, including links between users and RNs, users and BSs, and RNs and BSs, according to the increase order of the STR until the energy or throughput constraints can not be kept. We calculate and allocate the least number of sub-carriers required to each user for fulfilling traffic demand of users. After that, RNs are removed one by one according to the STR until any of cost, energy or throughput constraints can not be guaranteed. In the bottom-up algorithm, each user is connected to the closest BS and the least number of required sub-carriers is calculated. Then, the feasibility of current deployment is checked including all of the cost, energy and QoS constraints. If the placement is not a feasible solution, we deploy one more RN in a candidate location and connect users to the RN based on the STR until any constraint can not be kept.

3.4.2 Sub-Carrier and Traffic Over Rate

To design efficient heuristic algorithms for RNP-SA problem, the most important issue is to determine where to place the RNs and how to establish connections between users and RNs or BSs. In order to minimize the number of RNs, each RN should be able to serve as many users as possible to help relief the BSs' burden of energy and traffic demand. We propose a performance metric for designing efficient solution by jointly considering the energy consumption and users' traffic demand. We define our performance metric as follows:

Definition 1 Sub-carrier and Traffic over Rate of link (u, x) or (r, b):

$$
\begin{cases}
STR_{ux} = \frac{|S_{ux}|\gamma_u}{c_{ux}}, \forall u \in U, x \in B \cup R \\
STR_{rb} = \frac{|S_{rb}| \sum_{u \in \{u|(u,r) \in E\}} \gamma_u}{c_{rb}}, \forall r \in R, b \in B.
\end{cases}
\tag{3.12}
$$

Combining (3.12) with (3.2), the STR metric can be derived as

$$
\begin{cases}
STR_{ux} = \frac{\gamma_u}{W \log_2 (1+SNR_{ux})}, \forall u \in U, x \in B \cup R \\
STR_{rb} = \frac{\sum_{u \in \{u|(u,r) \in E\}} \gamma_u}{W \log_2 (1+SNR_{rb})}, \forall r \in R, b \in B.
\end{cases}
\tag{3.13}
$$

The *STR* is used to determine the minimal required active time for data transmission, which is the quotient of users' throughput requirement and achievable throughput of

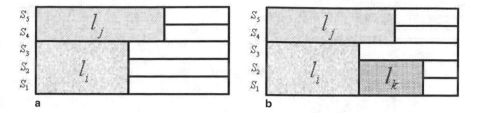

Fig. 3.2 Top-down and bottom-up sub-carrier allocation. **a** Top-down sub-carrier allocation. **b** Bottom-up sub-carrier allocation

the link with a single subcarrier. When a link has a smaller *STR*, it means that this link can achieve a higher rate and/or consume less energy compared with other links using the same number of sub-carriers.

3.4.3 RNP-SA with Top-Down Algorithm

We show the detail of our RNP-SA-t algorithm in Algorithm 1. We first deploy RNs in all candidate locations, establishing links between each user to a RN or BS and each RN to the BS in the increasing order of the STR. Since the total active time of all links in the same collision set should not exceed a unit time, we can obtain the least number of required sub-carriers, and then allocate each link with the least used available sub-carriers. An example of top-down sub-carrier allocation is shown in Fig. 3.2a.

Suppose that there are total 5 sub-carriers and the least required sub-carrier number of link l_i, link l_j and link l_k are 2, 3 and 2, respectively. The sub-carriers are allocated according to the sequence l_i, l_j and l_k as shown in Fig. 3.2. If either of energy or throughput is violated, no feasible solution will be returned. Otherwise, the STR of each RN is calculated and the RN with the least contribution of total STR will be deleted, i.e., the least difference of total STR with or without the RN. After that, we establish links between users and RN or BS according to the increasing order of STR, and then the summation of STR of all links will be calculated. Users of the deleted RN are connected to other RN or BS in the increasing order of STR, and the summation of STR of all links are calculated. We define each RN's contribution as the difference of total STR with or without the RN. The least contribution of RN is deleted as the node has least contribution to the network. Then, we repeatedly delete the least contribution of RN until any of cost, energy or QoS constraint is violated. The algorithm returns the number of deployed RNs in the last round when deleting $n + 1$-th RN will violate the constraints.

Algorithm 1 RNP-SA-t:RNP-SAwithtop-downalgorithm

Place RNs on all candidate locations;
while $R \neq \emptyset$ **do**
    Connect $(u, r, b) / (u, b)$ with $\min(STR_{urb}, STR_{ub})$,
    $\forall u \in U, r \in R, b \in B$;
    if $P_r^+ < P_r^- \vee P_b^+ < P_b^-$ **then**
        Connect u to the closest r or b
    end if
    Calculate the least required $|S_x|, \forall x \in V$;
    Top-down sub-carrier allocation;
    if $|R| = |R'| \wedge$ energy and/or QoS constraints can not be kept **then**
        Return no feasible solution;
    end if
    if Cost, energy and QoS constraints can be kept **then**
        Save the current topology;
        for all $r \in R$ **do**
            $U^* \leftarrow \{u : (u, r) \in E\}$
            $STR_1 \leftarrow \sum_{(u,r,b) \in E} STR_{urb}$;
            Delete r and disconnect its links;
            for all $u \in U^*$ **do**
                Add $(u, r, b) / (u, b)$ with $\min(STR_{urb}, STR_{ub})$;
                $STR_2 \leftarrow STR_2 + \min(STR_{urb}, STR_{ub})$;
            end for
            $STR^* \leftarrow STR_2 - STR_1$;
            Load the topology;
        end for
        Delete r with $\min(STR^*)$
    else
        Return relay node number of last loop;
    end if
end while
Return 0;

3.4.4 RNP-SA with Bottom-up Algorithm

To further decrease the time complexity of the heuristic algorithm, we propose the RNP-SA-b algorithm. We show the detail of our RNP-SA-b algorithm in Algorithm 2. In the bottom-up algorithm, all users are connected to their closest BSs at first, and then the least required number of sub-carriers under the cost, energy and throughput constraints is calculated. After that, the unused sub-carriers are assigned to each link according to the bottom-up sub-carrier allocation without time domain multiplexing. Therefore, a lower time complexity can be achieved by using bottom-up sub-carrier allocation comparing with the top-down sub-carrier allocation at the cost of reduced spectrum utilization. An example of the bottom-up sub-carrier allocation is shown in Fig. 3.2b. Since no more sub-carriers available, link l_k is not able to be scheduled for data transmission at this time. After allocating sub-carriers, the feasibility of current deployment is determined with the cost, energy and throughput constraints. If the cost constraint can not be kept, the program is stopped and the number of RNs is returned; If the energy constraint is violated, one more RN is deployed in the candidate location closest to the BS; If the throughput constraint can not be held, one more RN is placed in the candidate location with the heaviest traffic demand. After that, users are sorted by the STR of user-relay links in an ascending order. We establish the user-relay-BS link only if the cost constraint can not be kept, and the deployed RN can reduce the energy consumption of BSs with the fulfillment of users' throughput constraint. If either of the energy and throughput constraint can not be kept with the violation of cost constraint, no feasible solution will be returned by the algorithm.

3.4.5 Time Complexity Analysis of the RNP-SA-t/b Algorithms

In this subsection, time complexity of the proposed RNP-SA-t/b algorithms is analyzed under the worst case, which does not consider the cost constraint. The analysis is based on the assumption that the number of BSs in a two-tired wireless network is much smaller than that of RNs and their candidate locations. We can obtain the time complexity of RNP-SA-t/b as follows.

Lemma 1 *The time complexity of RNP-SA-t algorithm is $O(|S||L|(|U| + |L|)^2 + |U||L|^3)$.*

Proof For each round, construction of a new topology needs $O(|U||L|) + O(|L||B|)$. Since the number of BSs is much smaller than that of RNs and their candidate locations, we can get that $O(|U||L|) + O(|L||B|)$ is the same as $O(|U||L|)$. The sub-carrier allocation and feasibility checking by each node are conducted at the same time, which have the time complexity of $O(|S|(|U| + |L|))$. We need to check $O(|U| + |L|)$ nodes in total, which needs $O(|S|(|U| + |L|)^2)$ for algorithm to do sub-carrier allocation and feasibility checking. The energy constraint checking of each BS needs $O(|B|(|U| + |L|))$. To remove an RN, we need to investigate all candidate locations and its attached users, $O(|U||L|)$ and rearrange them $O(|L|)$, thus it takes $O(|U||L|^2)$ time. As there are $|L|$ rounds in the worst case and all steps are sequential, we can express the total complexity of RNP-SA-t algorithm as

$$O(|L|)[O(|U||L|) + O(|S|(|U| + |L|)^2)$$
$$+ O(|B|(|U| + |L|)) + O(|U||L|^2)]$$
$$= O(|S||L|(|U| + |L|)^2 + |U||L|^3).$$

$\square$

Algorithm2 RNP-SA-b:RNP-SAwithbottom-upalgorithm

Connect each user to the closest BS;
Calculate the least required $|S_x|$, $\forall x \in V$;
Bottom-up sub-carrier allocation;
if Cost, energy and QoS constraints can be kept **then**
    Return 0;
else
    while $P_x^+ < P_x^- \vee \sum_{(x,y) \in E} |S_{xy}| > |S|$ **do**
        if Cost constraint can not be kept **then**
            Return the number of placed RNs;
        else
            $b^* \leftarrow b_x$
            if $P_{b^*}^+ < P_{b^*}^-$ **then**
                $r \leftarrow z$ with $\min(D_{b^* z})$, $\forall z \in L \wedge SNR_{b^* z} \leq \beta$;
            else
                $r \leftarrow z$ with $\max(\gamma_z^l)$, $\forall z \in L \wedge SNR_{b^* z} \leq \beta$;
            end if
            $U^* \leftarrow$ sort users by STR_{ur} in increasing order;
            for all $u \in U^*$ **do**
                Replace (u, b) by connecting (u, r, b);
                if $P_r^+ < P_r^- \vee P_{b^*}^-$ is increased **then**
                    Replace (u, r, b) by connecting (u, b);
                end if
            end for
            Calculate the least required $|S_x|$, $\forall x \in V$;
            Bottom-up sub-carrier allocation;
            if $|R| = |R'| \wedge$ energy and/or QoS constraints can not be kept **then**
                Return no feasible solution;
            end if
            if Cost, energy and QoS constraints can be kept **then**
                Return the number of placed RNs;
            end if
         end if
        end while
end if

Lemma 2 *The time complexity of RNP-SA-b algorithm is* $O(|B||L|^2 + |L||U|\log|U| + |B||U||L|)$.

Proof The time complexity for determining which candidate location to place new added RN in each round is $O(|B||L|)$. After that, sorting users in increasing order of STR_{ur} needs $O(|U|\log|U|)$. it takes $O(|U|)$ to add users to the newly added RN, and the sub-carriers allocation for newly established links is $O(|U|)$. At last, we check the constraints for each BS which takes $O(|B|(|U| + |L|))$. As there are

$|L|$ rounds in the worst case and all steps are sequential, we can express the total complexity of RNP-SA-b algorithm as

$$O(|L|)[O(|B||L|) + O(|U|\log|U|)$$
$$+ O(|U|) + O(|U|) + O(|B|(|U| + |L|))]$$
$$= O(|B||L|^2 + |L||U|\log|U| + |B||U||L|).$$

$\square$

Therefore, RNP-SA-b algorithm has a lower time complexity than RNP-SA-t algorithm.

3.5   Simulation Results

In this section, we evaluate the performance of the RNP-SA-t/b algorithms by comparing with a traffic load oriented greedy algorithm. At first, we evaluate the performance of our algorithms by the minimal number of relay nodes required to deploy into the network. Then, the cost threshold is considered to limit the maximal number of relay node, and the average network lifetime is analyzed to evaluate the algorithm performance, where the network lifetime is the time duration from the beginning until one of the network nodes, either a BS or RN, drains out its energy and becomes out of service. After that we do our simulation under different network scenarios, e.g., diverse energy charging capabilities of candidate RNs at various locations, different traffic demands of each user, variable transmission powers, and different numbers of users or BSs.

3.5.1   Simulation Configurations

A two-tiered wireless network with 4 BSs, 150 wireless users, 50 candidate locations of RNs is set up within a $200 \, m \times 200 \, m$ region. Candidate locations and users are randomly distributed in the network, and BSs are evenly distributed. For all nodes, the same transmission power, $P^T = 0.5$ W, and receiving power, $P^R = 0.05$ W, are used to communicate with each other. Different candidate locations have various potential charging capabilities of BSs and RNs, and the energy charging rates of BSs/RNs are uniformly distributed over $[0.2, 0.4]$ W/ $[0.05, 0.1]$ W. There are total 50 sub-carriers in the network, and each sub-carrier has a bandwidth of 2 MHz. The background noise, the path loss exponent and interference signal threshold are set as $N = 10^{-4}$ W, 2 and 1, respectively. The traffic demand of users are randomly distributed over $[25, 55]$ Kbps, where downlink traffic demand is 9 times of that in the uplink. The simulation is conducted by using Java and Matlab, and each simulation is repeat by 1,000 times with different random seeds and the average values are computed for performance evaluation. We list the parameters used in the simulation as shown in Table 3.2.

Table 3.2 Table of parameters

Region size	200 m × 200 m
Number of BSs	4
Number of users	150
Number of candidate locations	50
Number of sub-carriers	50
Maximal network lifetime	200 time slots
Transmission power	0.5 W
Receiving power	0.05 W
Charging capability of BSs	[0.2, 0.4] W
Charging capability of RNs	[0.05, 0.1] W
Bandwidth of single sub-carrier	2 MHz
Pass loss exponent	2
Background noise	10^{-4} W
Interference signal threshold	1
Data-rate demand of users	[25, 55] Kbps
$\gamma_u^{dn}/\gamma_u^{up}, \forall u \in U$	9

3.5.2   Traffic Load Oriented Greedy Algorithm

A traffic load oriented greedy algorithm is considered as a benchmark for performance evaluation. The greedy algorithm first connects users to their closest BSs. Then, the least number of required sub-carriers is calculated for each user under the energy and throughput constraints. The greedy algorithm always places RNs into the candidate location with the heaviest traffic load, which is the candidate location has the highest sum of the traffic loads within the interference range. After that, the connections are established between users and their closest RNs, if the deployment cost, energy, and QoS constraints of RNs can be fulfilled and the deployment of the RN can help reduce BSs' energy consumption. We show the detail of traffic load oriented greedy algorithm in Algorithm 3.

Under the assumption that there is no cost threshold and the number of BSs is much smaller than the number of RNs or candidate locations, we can obtain the worst case time complexity of the traffic load oriented greedy algorithm.

Lemma 3 *The worst case time complexity of the traffic oriented greedy algorithm is $O(|L|^2(|L| + |U|))$.*

Proof In order to calculate the candidate location with heaviest traffic load, $O(|L|(|L| + |U|))$ is needed for each round. It takes $O(|U| \log |U|)$ and $O(|U|)$ to sort users and connect users to the newly added RN, respectively. The sub-carrier allocation for newly connected links needs $O(|U|)$. At last, the energy constraint checking takes $O(|B|(|U| + |L|))$ for all the BSs. As there are $|L|$ rounds in the worst case and all steps are sequential, we can express the total complexity of the traffic oriented greedy algorithm as

---

Algorithm3 Trafficloadorientedgreedyalgorithm

Connect each user to the closest BS;
Calculate the least required $|S_x|$, $\forall x \in V$;
Bottom-up sub-carrier allocation;
while Cost, energy and QoS constraints cannot be kept **do**
    $r \leftarrow z$ with max (γ_z^l), $\forall z \in L$;
    $b^* \leftarrow b$ with min (D_{rb}), $\forall b \in B$;
    Connect (r, b^*);
    $U^* \leftarrow$ sort users by D_{ur}, $\forall u \in U$;
    for all $u \in U^*$ **do**
        Replace (u, b^*) by connecting (u, r, b^*);
        if $P_r^+ < P_r^- \vee P_{b^*}^-$ is increased **then**
            Replace (u, r, b) by connecting (u, b);
        end if
    end for
    Calculate the least required $|S_x|$, $\forall x \in V$;
    Bottom-up sub-carrier allocation;
end while
return the number of placed RNs;

$$O(|L|)[O(|L|(|L| + |U|)) + O(|U| \log |U|)$$
$$+ O(U) + O(U) + O(|B|(|U| + |L|))]$$
$$= O(|L|^2(|L| + |U|)).$$

$\square$

Therefore, the traffic load oriented greedy algorithm and the RNP-SA-b algorithm have the same time complexity.

3.5.3   *Performance Evaluation*

In this subsection, the minimal required number of RNs are evaluated with various algorithms. The required number of RNs to fulfill traffic demand of users is shown in Fig. 3.3. We observe that more RNs are needed with the growth of users' traffic demand. As greedy algorithm only considers relieving the traffic burden of the BSs, it ignores the energy consumption. Our algorithms jointly consider the impact of energy sustainability and users' throughput requirement, which lead to better performance. We show the impact of variable energy charging capabilities of RNs in Fig. 3.4. It can be observed that the required number of RNs decreases with the growth of RNs' charging capability. This is because a higher capacity RN can provide services to more users, which helps to decrease the required number of RNs for a given number of users. We show the impacts of transmission power in Fig. 3.5. Normally, a higher

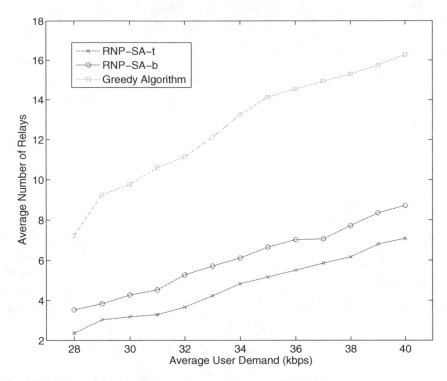

Fig. 3.3 Relay number of various user demand without cost threshold

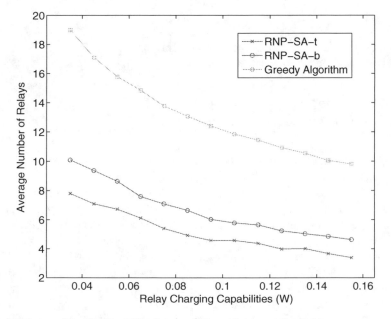

Fig. 3.4 Relay number of various charging capability without cost threshold

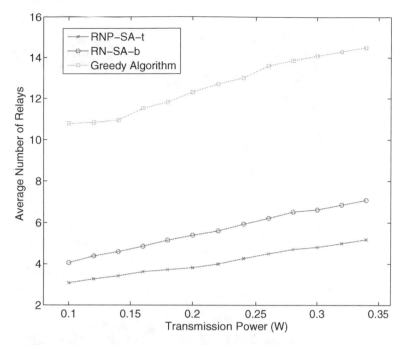

Fig. 3.5 Relay number of various transmission power without cost threshold

transmission power used by BSs to serve users can lead to more energy consumption, and thus more RNs are required to help release the burden of BSs. Overall, our proposed algorithms significantly outperform greedy algorithm, and the RNP-SA-t algorithm can achieve better performance than the RNP-SA-b algorithm with a higher time complexity. This is because the RNP-SA-t algorithm iteratively removes RNs based on all the candidate locations of the network topology information; while the RNP-SA-b algorithm can only utilize the current network topology information, which leads to a lower time complexity at the cost of a slightly lower performance comparing with the RNP-SA-t algorithm.

After that, the performance of network lifetime of various algorithms is evaluated based on a certain cost threshold. We show the network lifetime performance of various algorithms with different cost thresholds and the number of users in Figs. 3.6, 3.7 and 3.8. We observe that a higher cost threshold or a smaller number of users can achieve a longer network lifetime. This is because more RNs are allowed to be deployed, which can release more energy and throughput burden of BSs and improve the energy sustainability of the network. Normally, users' traffic demand deducts with the decrease of the number of users, which leads to a longer network lifetime. However, the greedy algorithm only focuses on the throughput constraint without considering energy constraint, which can not guarantee the sustainability of network. With the increase of cost threshold, the performance of network lifetime slightly increases in Fig. 3.8. By jointly considering the energy and throughput constraints, we propose two algorithms based on the STR performance metric for

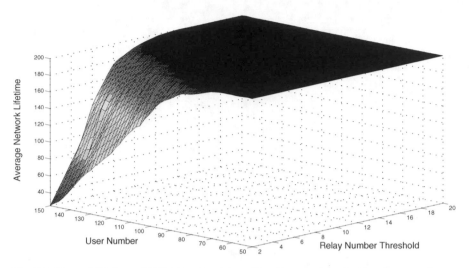

Fig. 3.6 Network lifetime of RNP-SA-t algorithm with various cost threshold and user number

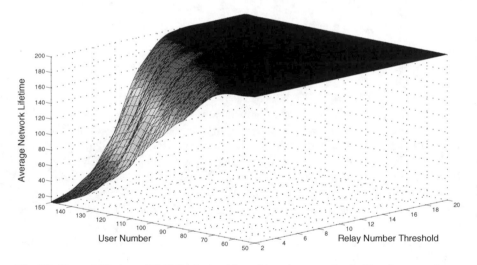

Fig. 3.7 Network lifetime of RNP-SA-b algorithm with various cost threshold and user number

RN deployment. We observe that the increase rate of network lifetime is much higher than that of the greedy algorithm in Figs. 3.6 and 3.7.

After that, we further investigate the performance of network lifetime based on various number of BSs in Fig. 3.9. We set the cost threshold as 5 RN, which means that the number of deployed RNs can not exceed 5. We can observe that the network lifetime extends significantly with the growth of BSs in Fig. 3.9. It can be found that the RNP-SA-t/b algorithms can significantly outperform the greedy algorithm. As greedy algorithm always places RNs in candidate locations with the heaviest

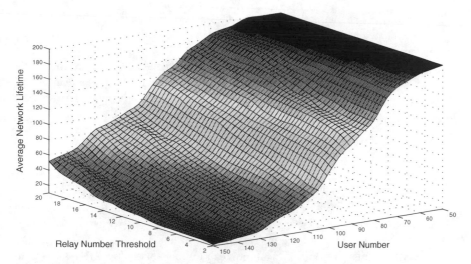

Fig. 3.8 Network lifetime of greedy algorithm with various cost threshold and user number

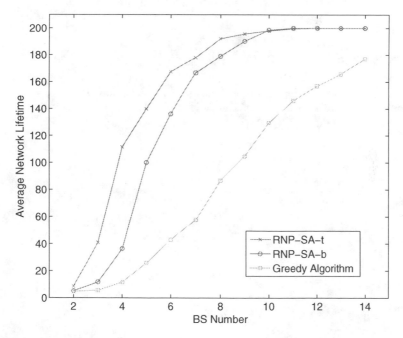

Fig. 3.9 Network lifetime of various number of BSs

traffic load, this strategy does not consider the energy efficiency especially when the heaviest load area is far away from the BSs.

In summary, the RNP-SA-t/b algorithms consider both the energy sustainability and users' throughput requirement, which can significantly outperform the traffic oriented greedy algorithm. With a higher time complexity to obtain the overall network topology information, the RNP-SA-t algorithm has slightly better performance than RNP-SA-b.

3.6 Summary

In this chapter, the joint problem of RN placement and sub-carrier allocation has been investigated under a two-tiered sustainable wireless network scenario. The RNP-SA problem has been formulated into an MINLP problem and two low-complexity heuristic algorithms are proposed to minimize the number of deployed RNs, subject to the energy, throughput and cost constraints. We have demonstrated that our proposed algorithms jointly consider the users' traffic demand and energy consumption, which can significantly outperform the traffic oriented greedy algorithm.

References

1. G. Lin and G. Xue, "Steiner tree problem with minimum number of steiner points and bounded edge-length," *Inf. Process. Lett*, vol. 69, no. 2, pp. 53–57, Jan. 1999.

Chapter 4
Analysis and Resource Management for Sustainable Wireless Networks

Using green energy sources to power wireless networks are believed as one of the efficient methods to mitigate the detrimental effects of conventional energy production or to enable deployment in off-grid locations. However, the dynamic characteristics of green energy sources in their availability and capacity have raised new challenges for related research issues like energy modeling and resource allocation in sustainable wireless networks. In this chapter, we focus on designing analytical energy modeling and admission control schemes to sustain the performance of sustainable wireless mesh network. specifically, our target is to maximize the energy sustainability of the network, or equivalently, to minimize the failure probability that the mesh APs deplete their energy and go out of service due to the unreliable energy supply. To this end, we use a $G/G/1(/N)$ queue with arbitrary patterns of energy charging and discharging to model the energy buffer of a green mesh AP. Based on diffusion approximation, the transient evolution of the queue length and the energy depletion duration are obtained. Then, an adaptive resource management scheme is proposed to balance the traffic load, while minimizing the failure probability at green mesh APs. We design a distributed admission control strategy to further enhance the network resource utilization and energy sustainability. Extensive simulation results show that our proposed schemes outperform some existing state-of-the-art solutions with considering the first and second order statistics of the energy charging and discharging processes at each green mesh AP.

This chapter is organized as follows. Introduction is presented in Sect. 4.1. The system model is introduced in Sect. 4.2 and an analytical framework is developed to study the transient buffer evolution in Sect. 4.3. Based on the buffer analysis, an adaptive resource management scheme is proposed in Sect. 4.4. Simulation results are given in Sect. 4.5, followed by concluding remarks in Sect. 4.6.

4.1 Introduction

There have been a lot of works addressing issues related to sustainable wireless communication networks. However, due to the complexity of green energy charging capability, most existing works study resource allocation by either assuming the

Z. Zheng et al., *Sustainable Wireless Networks,* SpringerBriefs in Computer Science,
DOI 10.1007/978-3-319-02469-1_4, © The Author(s) 2013

energy charging rate is known a *priori* or using an oversimplified model, e.g., the charging rate is uniformly distributed. In practice, the energy charging is a complicated and dynamic process, which is usually dependent on the charging capabilities of hardware, position, local weather and time. With the growing of users' traffic demand, it is critical to understand the impact of dynamic energy availability and analyze the charging and discharging processes of green energy to sustain the performance of green mesh wireless networks.

In this chapter, we focus on maximizing the energy sustainability of whole sustainable wireless mesh networks, such that the probability of green mesh APs depleting their energy and going out of service is minimized. Specifically, we develop an analytical model with dynamic charging and discharging processes to evaluate the instantaneous volume of buffered energy at green mesh APs. Based on analysis of the energy buffer, adaptive and optimal energy resource management are proposed to maximize the network-wide energy sustainability. Our main contributions are three-fold:

- *Analytical Model*: a generic analytical framework to study the transient evolution of the energy buffer is presented. Specifically, we model the energy buffer as a $G/G/1/\infty$ and $G/G/1/N$ queue with the general energy charging and discharging processes in infinite and finite energy buffer cases, respectively. Based on the first two statistical moments, i.e., the mean and variance, of the energy charging and discharging intervals, we apply the diffusion approximation to obtain closed-form distributions of the transient energy queue length and the energy depletion duration.
- *Adaptive Resource Management*: based on the developed analytical framework, we propose an adaptive resource management scheme to assure the energy sustainability of the network. In the proposal, traffic flows are distributively scheduled on a multi-hop path towards the minimal energy depletion probability of mesh APs. The proposed scheme is adaptive to the residual energy level at mesh APs and the dynamic traffic demands of flows.
- *Distributed Admission Control*: A distributed admission control strategy is deployed at mesh APs to strike a balance between high resource utilization and desirable energy sustainability requirement.

4.2   System Model

In this section, we present the energy model and network model configurations. A summary of the mathematical notations used in this chapter is given in Table 4.1.

4.2.1   Energy Model

In this chapter, a sustainable wireless mesh network with green mesh APs and a set of users is considered as shown in Fig. 4.1. For each green mesh AP, a battery and a

Table 4.1 The notations for Chap. 4.

$R_i(0)$	The initial battery energy of node i
$\mathcal{A}_i(t)$	The amount of energy charged over the time interval $[t-1,t]$ at node i
μ_a	The mean of the inter-charging intervals
v_a	The variance of the inter-charging intervals
$R_i(t)$	The residual energy of node i at time t
R_i^{max}	The maximum energy storage or the battery capacity of node i
R_i^{min}	The minimal residual energy level based on battery life and safety considerations
N_i	The maximum number of energy units that a node can use
e_r	Receiving and processing energy
$d_{i,j}$	The distance between nodes i and j
n	The path loss exponent
$e_{t_{i,j}}$	The minimal energy required for transmitting one bit over link $i \longrightarrow j$
$S_i(t)$	The total energy consumption of node i during the time interval $[t-1,t]$
$p_{i,j,k}(t)$	The traffic demand of flow k over link $i \longrightarrow j$ during $[t-1,t]$
μ_s	The mean of the energy inter-discharge interval
v_s	The variance of the energy inter-discharge interval
$v \in V$	Green mesh APs
$e \in E$	Wireless links between APs
w_v	The weight which represents the energy sustainability level of the node, i.e., how likely the node is to drop out by depleting all its energy to serve the traffic flows traversing it
$X(t)$	The continuous process to approximate the discrete buffer size $R(t)$
$G(t)$	A white Gaussian process with zero mean and unit variance
β	The mean of the change in $X(t)$
α	The variance of the change in $X(t)$
$\Phi(x)$	The standard normal integral
$\mathcal{D}(x_0)$	The energy depletion duration given the initial condition x_0, i.e., the duration from $X(0) = x_0$ until the moment when AP depletes energy
$\mathcal{P}(0; x_0)$	The probability of reaching the absorbing barrier starting from x_0
E	The approximation error function
$\hat{\mathcal{D}}$	The longest survival time of traffic flows
ε	An adjustable parameter which reflects the energy sustainability level
T	The survival time of a traffic flow

renewable energy supply are quipped to repeatedly charge energy from environment. The energy charging capabilities of different APs are dynamic over time, which is caused by various local environment including different solar radiation intensities or wind speeds. Thus, the energy charging process of each node is modeled as a continuous-time stochastic process, which has a arbitrary but stable distribution. The arrival events of the queue refer to the charged energy units, and we can use exponentially weighted moving average or other estimation approaches to estimate the mean and variance of the inter-charging intervals, μ_a and v_a. The harvested energy can be stored in the batteries, and the minimal residual energy level R_i^{min} is determined by considering battery life and safety, where $R_i^{min} \geq 0$. Hence, for any node i, its energy level is within the range $[R_i^{min}, R_i^{max}]$. To simplify the model, we consider $0 \leq R_i(t) \leq N_i$, where $N_i = R_i^{max} - R_i^{min}$, which is the maximum number of energy units that a node can use. We concern two scenarios, which are infinite energy buffer capacity, $N_i \to \infty$, and a limited energy buffer capacity, finite N_i.

The energy consumption consists of energy for receiving a packet, processing it, and transmitting it to the destination. We use a constant, e_r, to represent the receiving and processing energy, while the sender adjusts the transmission energy in order to guarantee a desired bit error rate. The energy path loss, $d_{i,j}^n$, is used to characterize the signal energy attenuates over a wireless channel [1, 2]. We can obtain the minimal required energy for transmitting one bit over link $i \longrightarrow j$ based on a given requirement of signal to noise ratio (SNR),

$$e_{t_{i,j}} \propto d_{i,j}^n. \tag{4.1}$$

We can express the total energy consumption of node i during the time interval $[t-1, t]$ as

$$S_i(t) = \sum_j \sum_k p_{i,j,k}(t)(e_r + e_{t_{i,j}}), \tag{4.2}$$

where $p_{i,j,k}(t)$ is the traffic demand of flow k over link $i \longrightarrow j$ during $[t-1, t]$. We can derive the residual energy of node i at time t as

$$R_i(t) = \min\{\max\{R_i(t-1) + A_i(t) - S_i(t), 0\}, N_i\}. \tag{4.3}$$

We use Poisson process to model the arrivals of traffic at each AP. Based on some estimation approaches, we can also estimate the values of μ_s and v_s.

4.2.2 Network Model

In this subsection, a distributed sustainable wireless mesh network is considered, which is composed by a set of stationary green mesh APs as shown in Fig. 4.1. Each local WLAN consists of a green mesh AP and wireless users, and green mesh AP responsible for communication between local WLAN and other WLANs. The network scenario is common, and one typical example is an office or home WLAN. In such kind of network scenario, a green mesh AP can be installed on the roof, such that it can harvest energy from natural resources, e.g., sun and wind, etc. Users located in a WLAN are allowed to communicate with local or remote users, where local AP helps both the communication within its coverage and the traffic from other APs toward the destination over the mesh backbone. We consider both intra-and inter-WLAN traffic demand in our system model. The green mesh APs utilize orthogonal frequency division multiplexing for data communication, thus the interference of intra- and inter-WLAN data transmission and reception is negligible. We suppose that the performance is limited by the energy instead of network bandwidth. Such kind of assumption is reasonable as network bandwidth is ample with modern wireless technologies and sustainable energy is not sufficient in general.

A direct graph $G = (V, E)$ for the mesh backbone network is generated, where green mesh APs are represented by vertices ($v \in V$) and wireless links between

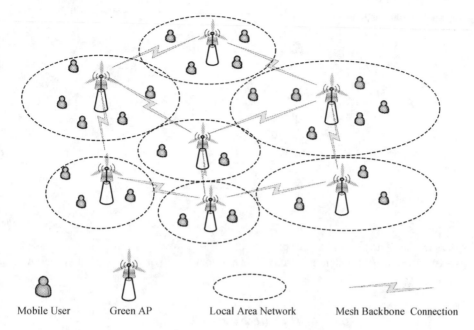

Fig. 4.1 WLAN Mesh Network

APs are represented by edges ($e \in E$). We use a weight w_v to represent the energy sustainability level of node v, i.e., how likely the node will use up its energy and break down. The transmission energy, $e_{t_{i \to j}}$, stands for the energy consumption of data transmission from node i to node j, which is a function of the link distance. The transmission energy of the forwarding links and traffic demands for the relay determines the energy consumption for relaying traffic.

4.3 Transient Queuing Analysis of Energy Buffer

In this section, a general framework is developed to analyze the characteristics of energy buffer. We focus on both infinite and finite battery storage cases and utilize a diffusion approximation to study the transient evolution of the buffer. $R(0) = x_0$ is the initial energy of a green mesh AP.[1] The energy buffer is used to store the harvested energy, and the battery capacity is N. Energy is discharged from the energy buffer when energy for traffic demand is needed. We show the evaluation of the energy buffer in Fig. 4.2. The dynamics of charging and discharging process may lead to the depletion of buffer energy, i.e., $R(t)$ reaches 0. In this case, the green mesh AP used up its energy and all the links associate with the AP are unreachable. Therefore,

---

[1] The node index in the notation is not included for simplicity in Sect. 4.2, e.g., the subscript i in $R_i(t)$.

Fig. 4.2 Evolution of an energy buffer

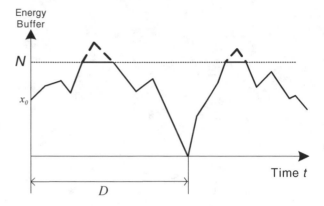

it is essential to minimize the probability of APs' energy depletion by properly distributing the traffic demands across the network based on energy level of each AP.

4.3.1 Queue Model of Infinite Energy Buffer

At first, we suppose that the capacity of battery is large enough to store all the harvested energy. We model the energy buffer as a $G/G/1/\infty$ queue, where μ_a (μ_s) and v_a (v_s) are its means and variances of inter-arrivals (inter-departures), respectively.

To analyze the $G/G/1/\infty$ queue [3], the diffusion approximation or Brownian motion approximation approach is used. We utilize a continuous process $X(t)$ to approximate the discrete buffer size $R(t)$, and the incremental change of $X(t)$ within a small period dt is normally distributed with mean βdt and variance αdt,

$$dX(t) = X(t + dt) - X(t) = \beta dt + G(t)\sqrt{\alpha dt}, \tag{4.4}$$

where $G(t)$ is a white Gaussian process with zero mean and unit variance. The mean and variance of the change in $X(t)$, β and α, represent the drift and diffusion coefficients, which are defined as

$$\beta = 1/\mu_a - 1/\mu_s \tag{4.5}$$

and

$$\alpha = v_a/\mu_a^3 + v_s/\mu_s^3. \tag{4.6}$$

Initially, the energy level is at $X(t = 0) = x_0$. We can derive the conditional probability density function (p.d.f.) of $X(t)$, which stands for the energy buffer size at time t ($t > 0$) as follows,

$$f(x, t; x_0)dx = Pr(x \le X(t) < x + dx | X(0) = x_0). \tag{4.7}$$

The p.d.f. of $X(t)$ satisfies the forward diffusion equation

$$\frac{\partial f(x,t;x_0)}{\partial t} = \frac{\alpha}{2}\frac{\partial^2 f(x,t;x_0)}{\partial x^2} - \beta\frac{\partial f(x,t;x_0)}{\partial x}, \tag{4.8}$$

under the boundary condition

$$X(t) \geq 0, t > 0. \tag{4.9}$$

Based on Eqs. 4.8, 4.9 and the method of images, we can obtain

$$f(x,t;x_0) = \frac{\partial}{\partial x}\left\{\Phi\left(\frac{x - x_0 - \beta t}{\sqrt{\alpha t}}\right)\right.$$
$$\left. - \exp\left\{\frac{2\beta x}{\alpha}\right\}\Phi\left(-\frac{x + x_0 + \beta t}{\sqrt{\alpha t}}\right)\right\}, \tag{4.10}$$

where $\Phi(x)$ is the standard normal integral defined as

$$\Phi(x) = \int_{-\infty}^{x}\frac{1}{\sqrt{2\pi}}\exp\left(-\frac{1}{2}z^2\right)dz. \tag{4.11}$$

The duration from $X(0) = x_0$ until the moment when AP depletes energy is called energy depletion duration, $\mathcal{D}(x_0)$, which can be expressed as

$$\mathcal{D}(x_0) = \inf\,(t \geq 0 | X(t) = 0, X(0) = x_0). \tag{4.12}$$

$\mathcal{D}(x_0)$ is also referred to as the first passage time of $X(t)$ from $x_0 > 0$ to 0.

We use the diffusion equation with the absorbing barrier at the origin to derive the density function of $\mathcal{D}$ as follows

$$f_{\mathcal{D}}(t;x_0) = \lim_{x\to 0}\left\{\frac{\alpha}{2}\frac{\partial f(x,t;x_0)}{\partial x} - \beta f(x,t;x_0)\right\}. \tag{4.13}$$

Based on the partial differential equation (Eq. 4.13), the conditional p.d.f. of $\mathcal{D}$ can be expressed as

$$f_{\mathcal{D}}(t;x_0) = \frac{x_0}{\sqrt{2\pi\alpha t^3}}\exp\left\{-\frac{(x_0 + \beta t)^2}{2\alpha t}\right\}. \tag{4.14}$$

Then, we can get the moment generation function of $\mathcal{D}$ as

$$f_{\mathcal{D}}^*(s;x_0) = \int_0^{\infty} e^{-st} f_{\mathcal{D}}(t;x_0)dt$$
$$= \exp\left\{-\frac{x_0}{\alpha}(\beta + \sqrt{\beta^2 + 2\alpha s})\right\}. \tag{4.15}$$

The probability $\mathcal{P}(0;x_0)$ of reaching the absorbing barrier starting from x_0 [3] can be calculated based on Eq. 4.15 as follows, when $s \to 0$.

$$\mathcal{P}(0; x_0) = \lim_{s \to 0} f_{\mathcal{D}}^*(s; x0)$$

$$= \begin{cases} 1, & \text{for } \beta < 0 \\ \exp\left(-\frac{2x_0\beta}{\alpha}\right), & \text{otherwise.} \end{cases} \quad (4.16)$$

When the energy discharging rat is larger than or equal to the energy harvesting range, i.e., $1/\mu_s \geq 1/\mu_a$, Eq. 4.16 shows that the probability of energy buffer depletes is 1. We also need to discuss the case that the mean of energy charging rate is larger than the mean of energy discharging rate, because the variance of energy charging and discharging processes may lead to energy depletion. In this case, the probability of energy depletion is determined by several parameters including the mean and variance of energy charging and discharging processes, and the initial energy level x_0,.

The mean and variance of $\mathcal{D}$ for $\beta \neq 0$ can be obtained based on the differentiation of (4.15) w.r.t. s,

$$E[\mathcal{D}; x_0] = x_0/|\beta|, \quad (4.17)$$

$$V[\mathcal{D}; x_0] = x_0\alpha/|\beta|^3. \quad (4.18)$$

4.3.2 Queue Model of Finite Energy Buffer

After analyze unlimited battery charging and discharging process, we consider more realistic case, i.e., a green mesh AP has a limited battery capacity, and the energy buffer is modeled as a $G/G/1/N$ queue.

Similarly, we can obtain that the conditional p.d.f. of the buffer size satisfies the forward diffusion equation based on the initial condition and the buffer size N.

$$\frac{\partial f(x, t; x_0, N)}{\partial t}$$

$$= \frac{\alpha}{2} \frac{\partial^2 f(x, t; x_0, N)}{\partial x^2} - \beta \frac{\partial f(x, t; x_0, N)}{\partial x} \quad (4.19)$$

$$+ \mu_a P_0(t; x_0, N)\delta(x - 1)$$

$$+ \mu_s P_N(t; x_0, N)\delta(x - N + 1),$$

where $P_0(t)$ and $P_N(t)$ are the probability mass functions at the boundary $x = 0$ or $x = N$ at time t, which can be expressed as follows based on the initial value x_0:

$$P_0(t; x_0, N) = Pr[X(t) = 0 | X(0) = x_0] \quad (4.20)$$

$$P_N(t; x_0, N) = Pr[X(t) = N | X(0) = x_0]. \quad (4.21)$$

If one of the boundaries $x = 0$ or $x = N$ has been reached by $X(t)$, a random interval will be held with mean $1/\mu_a$ and $1/\mu_s$, respectively. We also need to ensure that the probability mass functions $P_0(t; x_0, N)$ and $P_N(t; x_0, N)$ can satisfy the following equations:

$$\frac{dP_0(t; x_0, N)}{dt} = -\mu_a P_0(t; x_0, N) \tag{4.22}$$

$$+ \lim_{x \to 0} \left[\frac{\alpha}{2} \frac{\partial f(x, t; x_0, N)}{\partial x} - \beta f(x, t; x_0, N) \right]$$

and

$$\frac{dP_N(t; x_0, N)}{dt} = -\mu_N P_N(t; x_0, N)$$

$$+ \lim_{x \to N} \left[-\frac{\alpha}{2} \frac{\partial f(x, t; x_0, N)}{\partial x} + \beta f(x, t; x_0, N) \right], \tag{4.23}$$

subject to the initial condition

$$f(x, t; x_0, N) = \delta(x - x_0) \, 0 < x < \infty, \tag{4.24}$$

and boundary conditions

$$\lim_{x \to 0} f(x, t; x_0, N) = 0 \, t > 0 \tag{4.25}$$

$$\lim_{x \to N} f(x, t; x_0, N) = 0 \, t > 0. \tag{4.26}$$

Based on a doubly infinite system of images [3], the transient solution of the probability function can be derived as

$$f(x, t; x_0, N) = \frac{1}{\sqrt{2\pi \alpha t}} \sum_{n=-\infty}^{\infty} (A - B), \quad t > 0,$$

where

$$\begin{cases} A = \exp \left(\frac{\beta 2nN}{\alpha} - \frac{(x - x_0 - 2nN - \beta t)^2}{2\alpha t} \right), \\ B = \exp \left(\frac{\beta(-2x_0 - 2nN)}{\alpha} - \frac{(x + x_0 + 2nN - \beta t)^2}{2\alpha t} \right). \end{cases} \tag{4.27}$$

As $\sum_i P_i = 1$, we can obtain

$$\int_0^N f(x, t; x_0, N) dx + P_0(t; x_0, N) + P_N(t; x_0, N) = 1. \tag{4.28}$$

Similar to Eq. 4.13, we should guarantee the density function of the first passage time in the finite buffer case satisfies

$$f_D(t; x_0, N) = \lim_{x \to 0} \left[\frac{\alpha}{2} \frac{\partial f(x, t; x_0, N)}{\partial x} - \beta f(x, t; x_0, N) \right]. \tag{4.29}$$

The Laplace transform of the density function $f_D(t; x0, N)$ with a finite energy capacity N can be expressed as

$$f_D^*(s; x_0, N) = \exp\left\{-\frac{\beta}{\alpha}x_0\right\} \frac{\sinh\left[C(N - x_0)\right] - D}{\sinh\left(CN\right) - D} \tag{4.30}$$

where

$$\begin{cases} C = \frac{\sqrt{2\alpha s + \beta^2}}{\alpha}, \\ D = \frac{\mu_a}{\mu_a + s} \exp\{\beta/\alpha\} \sinh\left[C(N - 1)\right]. \end{cases}$$

We can approximate the inversion of $f_D^*(s; x_0, N)$ as

$$\begin{aligned}
& f_D(t; x0, N) \\
&= \frac{1}{2t} \exp\left\{\frac{E}{2}\right\} \Re\left(f_D^*\left(\frac{E}{2t}; x_0, N\right)\right) \\
& + \frac{1}{t} \exp\left\{\frac{E}{2}\right\} \sum_{k=1}^{\infty} (-1)^k \Re\left(f_D^*\left(\frac{E + 2k\pi i}{2t}; x_0, N\right)\right),
\end{aligned} \tag{4.31}$$

where E stands for the approximation error function. It is used to ensure that the error is bounded by $\frac{\exp\{-E\}}{1-\exp\{-E\}}$ [4]. For example, the approximation error can be guaranteed to be smaller than $0.99 \cdot 10^{-5}$ if we set $E = 5\ln(10)$.

After that we can drive the first moment of D based on the differentiation of (4.30) w.r.t. s,

$$\begin{aligned}
& E[D; x_0, N] \\
&= \begin{cases} -\frac{x_0}{\beta} + (\mu_a + \frac{1}{\beta})\frac{\exp\{-\frac{2\beta}{\alpha}(N-x_0)\} - \exp\{-\frac{2\beta}{\alpha}N\}}{1 - \exp\{-\frac{2\beta}{\alpha}\}}, & \beta \neq 0, \\ x_0(\mu_a + \frac{2\,N - x_0 - 1}{\alpha}), & \beta = 0. \end{cases}
\end{aligned} \tag{4.32}$$

Solving Eqs. (4.19)–(4.23) at the stationary state when $\lim_{t\to\infty} P_0(t; x_0, N) = P_0$, $\lim_{t\to\infty} P_N(t; x_0, N) = P_N$, and $\lim_{t\to\infty} f(x, t; x_0, N) = f(x; x_0, N)$,

$$\begin{aligned}
& \frac{\alpha}{2} \frac{\partial^2 f(x, t; x_0, N)}{\partial x^2} - \beta \frac{\partial f(x, t; x_0, N)}{\partial x} \\
&= -\mu_0 P_0 \delta(x - 1) - \mu_N P_N \delta(x - N + 1),
\end{aligned} \tag{4.33}$$

$$\lim_{x\to 0}\left[\frac{\alpha}{2} \frac{\partial p(x, t; x_0, N)}{\partial x} - \beta p(x, t; x_0, N)\right] = \mu_0 P_0, \tag{4.34}$$

$$\lim_{x\to N}\left[\frac{\alpha}{2} \frac{\partial p(x, t; x_0, N)}{\partial x} - \beta p(x, t; x_0, N)\right] = -\mu_N P_N, \tag{4.35}$$

The steady state probability function of the energy buffer length [5] can be expressed as

$$
f(x; x_0, N)
$$

$$
= \begin{cases} -\frac{\mu_a P_0}{\beta}[1 - e^{rx}], & 0 \le x \le 1 \\ -\frac{\mu_a P_0}{\beta}[e^{-r} - 1]e^{rx}, & 1 \le x \le N - 1 \\ -\frac{\mu_a P_0}{\beta}[e^{r(x-N)} - 1]e^{r(N-1)}, & N - 1 \le x \le N, \end{cases} \tag{4.36}
$$

where $r = (2\beta)/\alpha$.

We can drive the probability that an AP depletes energy in the finite buffer case as

$$
\mathcal{P}(0; x_0, N)
$$

$$
= \begin{cases} (1 + \frac{\mu_s}{\mu_a}e^{r(N-1)} + \frac{\mu_s}{\mu_a - \mu_s}[1 - e^{r(N-1)}])^{-1}, & \beta \ne 0, \\ \frac{1}{2}(1 + \frac{N-1}{v_a^2/\mu_a^2 + v_s^2/\mu_s^2})^{-1}, & \beta = 0. \end{cases} \tag{4.37}
$$

In the finite energy buffer case, we can observe from Eq. 4.37 that the variance of the charging and discharging processes may lead to a certain probability of energy depletion according to the first and second moments of the sojourn times at the boundary conditions, which is regardless of whether $\beta \ge 0$ or $\beta < 0$. We found that a smaller battery capacity N/the energy charging rate or a larger variance in either charging or discharging result in a higher energy depletion probability $\mathcal{P}(0; x_0, N)$.

4.4   Adaptive Resource Management

In this section, an energy-aware adaptive resource management framework is developed based on the transient energy buffers of green mesh APs. Our target is to minimize the energy depletion probability, i.e., green mesh APs used up their energy to serve traffic demands, such that the network can be sustained. By jointly considering parameters including the transient energy level, energy charging capability, and existing traffic demand at each AP, a path selection metric is designed to distribute traffic flows for balancing the traffic load. We also present a distributed admission control strategy to further sustain the network.

4.4.1   Relay Path Selection

Normally, multi-hop transmission is needed for users' traffic to reach its destination in WLAN mesh networks as shown in Fig. 4.1. It is essential to schedule traffic flows over relay paths in order to minimize the energy depletion probability, such that the network sustainability and connectivity can be guaranteed. The scheduling should be updated periodically to dynamically track the charging capability and traffic demands.

Our relay path selection design is based on the analysis in Sect. 4.3. At first, a request is broadcasted by a source user, which has the information of the destination and the estimated first and second order statistics of the energy consumption of the traffic flow. The destination AP calculates the energy depletion probability $\mathcal{P}(0; x_0)$ (Eq. 4.16) or $\mathcal{P}(0; x_0, N)$ (Eq. 4.37) according to the current energy level and the accumulated traffic demands, when it receives the request from the source user. After that, the weight of destination AP is updated as

$$w_v = \mathcal{P}(0; x_0) \quad \text{or} \quad \mathcal{P}(0; x_0, N). \tag{4.38}$$

The destination responses the source user by sending a reply message with its weight. The green mesh AP updates its weight when the reply message is received. The source AP computes its own $\mathcal{P}(0; x_0)$ with the accumulated load energy consumption over each link and selects the path for data transmission. For instance, the path with the minimal sum of the energy depletion probability (MEDP), i.e., the path with the minimal $\sum w_v$, may be selected, such that the overall network sustainability can be guaranteed; the path along which the maximum value of w_v is minimized, i.e., the path with Min Max w_v, may be selected in order to sustain the least sustainable AP. We use the shortest path algorithm to update the weight, where the number of hops is replaced by the path selection metric, i.e., $\sum w_v$ or Max w_v.

In the infinite buffer capacity case, it is possible that $\mathcal{P}(0; x_0)$ increases to 1 when $\beta \leq 0$,, which means that the energy of the AP will eventually be used up by relaying the traffic flow. Thus, the AP needs to evaluate the current energy level and make sure that the energy is sufficient to fulfill the flow demand within a finite duration. We use T to represent the survival time of the flow in the following T time slots. Based on the survival time, the AP calculates the probability that the energy is used up before T expires, which can be expressed as

$$F_{\mathcal{D}}(T; x_0) = Pr(\mathcal{D} \leq T) = \int_0^T f_{\mathcal{D}}(t; x_0) dt. \tag{4.39}$$

In the infinite buffer case, the weight of the destination AP is updated according to the flow request as

$$w_v = \mathcal{P}(0; x_0) + I(\beta \leq 0) F_{\mathcal{D}}(T; x_0), \tag{4.40}$$

where $I(\cdot) = 1$ if condition $(\cdot)$ is true and 0 otherwise. In a network of queues, the variance of traffic may be absorbed by the relaying AP's buffer, which may also vary the output traffic characteristics. APs update the traffic parameters and repeat the path selection process mentioned before if the estimated traffic demand changes during time duration T. If the hop green mesh AP depletes its energy, its previous hop green mesh AP re-sends a packet after a random period, which may also change the energy consumption statistics of the ongoing flow. If $\mathcal{P}(0; x_0)$ is large, the energy consumption statistics may vary hop by hop, and the energy consumption statistics and $\mathcal{P}(0; x_0)$ should be updated by green mesh APs. However, we can guarantee the sustainability of green mesh APs by minimizing the energy depletion probability and ensuring a sufficiently large depletion duration.

4.4.2 Admission Control

It is essential to apply admission control to provide satisfactory quality of services for users, especially when the network resources is limited. Normally, the admission control needs to consider the trade off between the resource utilization and quality of service provisioning. For example, the network can achieve a higher network throughput by admitting more users to the network, meanwhile the network resources are consumed faster which may lead to cease operation of some devices. Hence, an effective admission control strategy is critical to guarantee efficient resource utilization under the constraints of energy sustainability and throughput requirement.

We focus on providing qualified services to admitted users by guaranteeing a sufficient duration of energy depletion in a stochastic manner. For example, a required service quality should be ensured, i.e., the probability that $\mathcal{D}$ is larger than the longest survival time of traffic flows $\hat{\mathcal{D}}$ should exceed a threshold, which can be expressed as

$$Pr(\mathcal{D} \le \hat{\mathcal{D}}) = \int_0^{\hat{\mathcal{D}}} f_{\mathcal{D}}(t; x_0)dt < \varepsilon, \qquad (4.41)$$

where the parameter $0 < \varepsilon \ll 1$ stands for the energy sustainability level. The threshold ε denotes the strictness of energy sustainability constraint for admitting a new flow. Thus, green mesh APs determine their availability for transmission relay based on the estimated flow statistics, which means a green mesh AP only responses to a request when the energy sustainability condition in Eq. 4.41 can be fulfilled. If the source AP cannot successfully generate a relay path to the destination due to one or more APs depleting the energy, the flow request will be rejected by the source AP. The network can achieve a satisfactory sustainability by upper bounding the energy depletion probability.

4.5 Simulation Results

In this section, extensive simulations are conducted to analyze the energy buffer and evaluate the performance of the proposed resource management schemes, based on a discrete time event-driven simulator coded in C++.

4.5.1 Simulation Setup

In the simulation, a WLAN mesh network with ten APs and multiple mobile users is generated as shown in Fig. 4.1. We randomly select the distance between adjacent APs in $[R_0, 3R_0]$. The communication radius of each AP is set as R_0, and ten groups of users are randomly distributed in the coverage of APs. For each green mesh AP, the energy charging intervals are randomly selected from $\mathbf{t_a} = \{1, 2, 3, 4\}$ (in unit of time slots) with a given probability distribution $\mathbf{p_{t_a}}$, e.g., if $\mathbf{p_{t_a}} = \{0.3, 0.3, 0.2, 0.2\}$, the mean and variance of the charging intervals' duration are $\mu_a = 2.3$ and $\nu_a = 1.21$,

Fig. 4.3 $\mathcal{P}(0; x_0)$ (Infinite buffer, $\mu_a = 2.3$, $v_a = 1.21$, $\mu_s = 2.33$, $v_s = 5.44$)

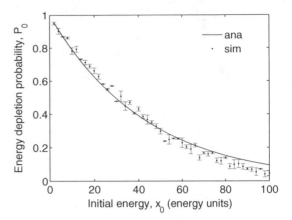

respectively. The energy consumption of each AP is up to the traffic demand of users. A number of flows are randomly generated with any two users as source and destination in each round of simulation. The inter-packet arrivals of a flow are exponentially distributed with mean $\mu_s = 14$. The energy consumption of each packet consists of reception and transmission. For intra-WLAN traffic, we set the energy consumption of the transmission between the AP and any user within its coverage R_0 as a constant $e_0 = 1$ energy unit. For inter-WLAN traffic, the energy consumption of the transmission between green mesh APs depends on the link distance R. R can be obtained by $max\{1, (R/R_0)^n\}e_0$ where n is selected from $\{1, 2, 3, 4\}$. The energy consumption of signaling exchange is ignored as relatively negligible low energy consumption compared to the traffic demand. Each simulation is run by 10 times, which contains 1,000 runs with different random seeds. We use 95 % confidence bars to calculate the average results.

4.5.2 Energy Buffer

At first, we evaluate the energy depletion probability under the case of both infinite and finite buffer. We add six flows to transmit over a link with distance R_0 one by one and analyze the buffer depletion to study the relationship between the charging capability and traffic load consumption. For six flows, we know that the departure rate is smaller than the energy arrival rate and $\beta > 0$. We do not set a survival time for these flows in order to analyze the buffer evolution. In Fig. 4.3, we record each time that the energy buffer of node A reaches 0, and divide the number of runs when the simulation runs 6,000 time slots by the total number runs of 1,000 to calculate $P(0; x_0)$. We can observe from Fig. 4.3 that $\mathcal{P}(0; x_0)$ decreases with the initial energy x_0. With limited duration (6,000 time slots), we get conservative simulation results, which is slightly lower than the analytical results (It converges when time goes to infinity). When the energy buffer is finite as shown in Fig. 4.4, the energy depletion probability decreases with the growth of energy buffer capacity.

Fig. 4.4 $\mathcal{P}(0; x_0, N)$ (Finite buffer, $\mu_a = 2.3$, $v_a = 1.21$, $\mu_s = 2.33$, $v_s = 5.44$)

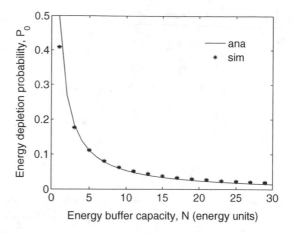

Fig. 4.5 CDF of $\mathcal{D}$ (Infinite buffer, $\mu_a = 2.3$, $v_a = 1.21$, $\mu_s = 1.16$, $v_s = 1.36$)

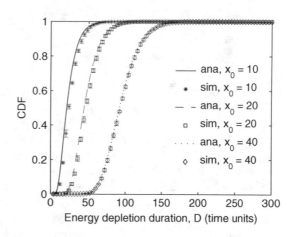

In Fig. 4.5, the cumulative distribution function (CDF) of the energy depletion duration is plotted under the case of infinite energy buffer. Six more flows to transmit over a link with distance R_0 are added and the first passage time is calculated to effectively obtain the CDF of $\mathcal{D}$. As the energy discharging rate is higher than the energy charging rate for 12 flows, the energy buffer will be stable with $\beta < 0$ and the energy will be used up eventually. We can observe from Fig. 4.5 that a smaller x_0 means AP is more likely to deplete its energy in the near future, which makes the CDF curve shift to the left. In Fig. 4.6, the CDF of the energy depletion duration is plotted under the case of finite energy buffer. Initially, the energy buffer is full, which means that $x_0 = N$. If there is a large energy buffer capacity and initial buffer size, it is more likely that an AP depletes its energy soon, which leads to the CDF curve shifts to the right.

Fig. 4.6 CDF of $\mathcal{D}$ (Finite
buffer, $\mu_a = 2.3, v_a = 1.21$,
$\mu_s = 2.33, v_s = 5.44$)

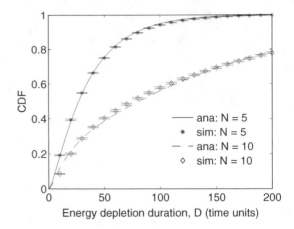

4.5.3   Sustainable Network Performance

In this subsection, we evaluate the network lifetime under various parameters, where the network lifetime is the maximal duration from the start of network until one of the APs depletes its energy. A heterogeneous network is considered, where green APs with different locations have diverse charging capabilities by using various probability vectors $\mathbf{p}_{t_a}$.

The proposed MEDP with path selection metric of $\sum w_v$ and Max w_v is compared with two other schemes, i.e., the minimum energy (ME) scheme [6], and the minimum path recovery time (MPRT) scheme [7]. The ME scheme selects a relay path with the minimal total energy consumption along the path; The MPRT algorithm focuses on minimizing cumulative recovery time of the selected path, such that the recovery rate of consumed energy could be maximized. Therefore, MPRT normally chooses the path with a higher charging rate comparing with the ME scheme.

We compare the three path selection schemes in Figs. 4.7 and 4.8. We found that the MPRT and MEDP outperform a lot comparing to ME without considering the energy charging capability in the path selection. The MEDP can achieve a better performance comparing with MPRT as MPRT only considers the charging capability of green mesh APs, while the MEDP jointly concerns traffic demands, charging capability and variations in both charging and discharging processes. The proposed MEDP focuses on maximizing the overall network sustainability by using the metric $\sum w_v$, which can also guarantee the worst case performance of network sustainability. Both metrics in the MEDP can achieve similar performance. It is found that the growth of the initial energy buffer x_0 and the energy buffer capacity N can enhance the network lifetime as shown in Figs. 4.7 and 4.8. The proposed MEDP scheme can significantly extend the network life time in both finite buffer and infinite buffer cases with a given initial buffer x_0 or the buffer capacity N.

We compare the network lifetime under the case with and without CAC schemes applied in Fig. 4.9. Without CAC, the performance of network lifetime is degraded significantly with the growth of traffic loads caused by increasing admitted flows in

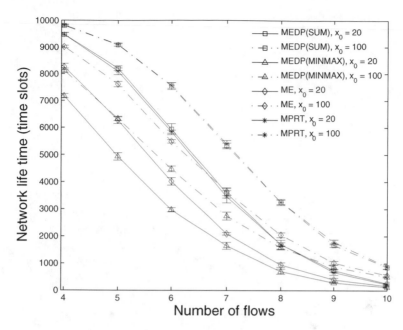

Fig. 4.7 Path selection comparison (Infinite buffer)

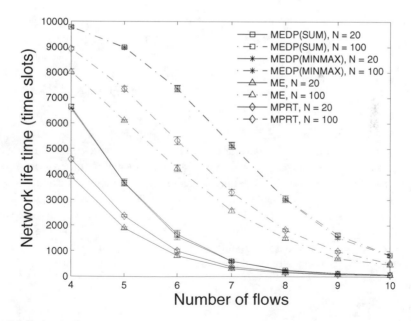

Fig. 4.8 Path selection comparison (Finite buffer)

Fig. 4.9 Network life time w/wo CAC (Infinite buffer)

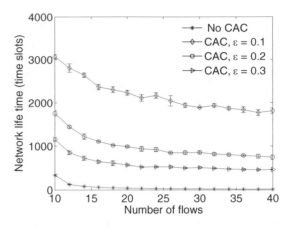

Fig. 4.10 Number of admitted flows (Infinite buffer)

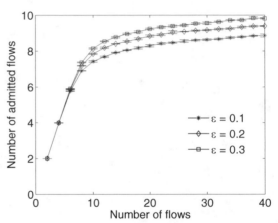

the network. In this simulation, we allow the first six flows to join the network and reject some following flows according to whether the energy condition of APs can sustain their traffic demands or not. We show the number of admitted flows under various value of ε in Fig. 4.10. A smaller ε leads to a limited network throughput with a stricter admission condition. We can observe that one more flow can be admitted when ε is increased from 0.1 to 0.2. However, the increased flow may deplete some APs' energy and the network lifetime may be decreased a lot as shown in Fig. 4.9. Hence, a smaller ε leads to less possibility of energy depletion, which can enhance the overall network connectivity and service provisioning to the existing flows.

4.6   Summary

In this chapter, we try to maximize the energy sustainability of green mesh APs in order to sustain the network performance. Specifically, a generic analytical model has been developed to characterize the transient evolution of an energy buffer. After that, we have derived the closed-form solutions of energy buffer analysis including the energy depletion probability and energy depletion duration. Based on the analysis results, we have proposed an adaptive resource management framework for relay path selection and admission control. Our extensive simulation results show that the network sustainable performance can be significantly improved by mitigating the energy depletion probability and ensuring a large energy depletion duration of green mesh APs.

References

1. X. Cheng, C. X. Wang, H. Wang, X. Gao, X. H. You, D. Yuan, B. Ai, Q. Huo, L. Y. Song, and B. L. Jiao, "Cooperative MIMO channel modeling and multi-link spatial correlation properties," *IEEE Journal on Selected Areas in Communications*, vol. 30, no. 2, pp. 388–396, Feb. 2012.
2. C. X. Wang, X. Hong, X. Ge, X. Cheng, G. Zhang, and J. Thompson, "Cooperative MIMO channel models: A survey," *IEEE Communications Magazine*, vol. 48, no. 2, pp. 80–87, Feb. 2010.
3. D. R. Cox and H. D. Miller, *The Theory of Stochastic Processes*. Chapman and Hall Ltd. London, 1965.
4. M. Parlar, *Interactive Operations Research with Maple: Methods and Models*. Birkhauser, Boston, Jul. 2000.
5. E. Gelenbe, "On approximate computer system models," *J. ACM*, vol. 22, no. 2, pp. 261–269, 1975.
6. A. Srinivas and E. Modiano, "Minimum energy disjoint path routing in wireless ad-hoc networks," in *MobiCom*. New York, NY, USA: ACM, 2003, pp. 122–133.
7. E. Lattanzi, E. Regini, A. Acquaviva, and A. Bogliolo, "Energetic sustainability of routing algorithms for energy-harvesting wireless sensor networks," *Computer Communications*, vol. 30, no. 14-15, pp. 2976–2986, Oct. 2007.

Chapter 5
Conclusions and Future Directions

In this brief, we aim at maximizing the network sustainability of wireless communication networks with green energy. The main content of this brief is shown as follows:

- We have investigated the green techniques and previous works related to sustainable wireless networks. Several research issues have been discussed including network planning, energy modeling and resource allocation. Moreover, the motivations and challenges related to these research issues have been highlighted.
- We have addressed the relay node placement and sub-carrier allocation (RNP-SA) issues, which have been jointly formulated into a mixed integer non-linear programming problem. Two low-complexity heuristic algorithms, namely RNP-SA with top-down/bottom-up algorithms (RNP-SA-t/b), have been presented to solve the RNP-SA problem. We have analyzed the time complexity of proposed algorithms, and our algorithms have shown good performance by numerical and simulation results.
- We have analyzed the characteristics of green energy by developing a generic analytical model. The closed-form solutions for the energy depletion probability and energy depletion duration have been derived. Based on the analysis, an adaptive resource management framework for relay path selection and admission control has been proposed. Extensive simulation results demonstrate that the network sustainability can be improved by mitigating the energy depletion probability and ensuring a large energy depletion duration of mesh APs.

The dynamic energy harvest and traffic demand lead to many new challenging research issues under the scenario of sustainable wireless networks [1–3]. We close this chapter by presenting our future works as research directions in this field.

- To construct a sustainable wireless communication network, the sustainable wireless devices should be carefully deployed into the network area based on users' demand, different charging capabilities at various locations, and the remaining energy in buffers. The variation and dynamic energy charging and energy consumption should be considered. In addition, mobility models should be considered for different network scenarios [4–6]. For instance, green-energy-powered roadside units deployment in VANET is a typical application scenario, and further

Z. Zheng et al., *Sustainable Wireless Networks,* SpringerBriefs in Computer Science, DOI 10.1007/978-3-319-02469-1_5, © The Author(s) 2013

consideration on the uniqueness of VANET mobility (e.g., highly dynamic but predictable) is indispensable for efficient green roadside unit deployment.

- Usually, the QoS requirement of users is various and varying based on capacities and usage of their terminal devices [7–9]. Moreover, the dynamic characteristics of green energy lead to a variety of charging capabilities for green network devices, which have significant impact on the allocation of network resources. Thus, to provision satisfactory services to each user, efficient resource allocation methods, e.g., sub-carrier allocation, power control, scheduling, etc., are required to allocate the limited network resources according to diverse requirement of users. A resource allocation scheme should be considered to allocate the sub-carriers, to adjust transmission power, and to schedule the active time for each link, such that the network sustainability and dynamic user requirement can be guaranteed with harvested energy.
- Routing scheme design is a very important issue in sustainable wireless networks [10, 11]. The uneven energy consumption of users and charging capabilities of BSs may make some routers overdraw its energy buffer and may lead to network system breaking down. Thus, to guarantee the normal operations of green wireless communication networks, routing scheme should be carefully designed to prevent the transmission going through the node which has not sufficient remaining energy in its energy buffer. The proposed routing scheme should be able to determine the proper route for each transmission, such that the QoS requirement of each transmission can be guaranteed and the remaining energy in energy buffer can support the transmissions.

References

1. J. Wu, "Green wireless communications: from concept to reality," *IEEE Wireless Communications*, vol. 19, no. 4, pp. 4–5, Aug. 2012.
2. L. C. Wang and S. Rangapillai, "A survey on green 5G cellular networks," in *Signal Processing and Communications (SPCOM)*, Bangalore, IN, 22–25 Jul. 2012, pp. 1–5.
3. S. Yeh, "Green 4G communications: Renewable-energy-based architectures and protocols," in *Global Mobile Congress (GMC)*, Shanghai, CN, 18–19 Oct. 2010, pp. 1–5.
4. X. Lin, R. K. Ganti, P. J. Fleming, and J. G. Andrews, "Towards understanding the fundamentals of mobility in cellular networks," *IEEE Transactions on Wireless Communications*, vol. 12, no. 4, pp. 1686–1698, Apr. 2013.
5. X. Wang, X. Lin, Q. Wang, and W. Luan, "Mobility increases the connectivity of wireless networks," *IEEE/ACM Transactions on Networking*, vol. 21, no. 2, pp. 440–454, Apr. 2013.
6. Q. Dong and W. Dargie, "A survey on mobility and mobility-aware MAC protocols in wireless sensor networks," *IEEE Communications Surveys and Tutorials*, vol. 15, no. 1, pp. 88–100, Feb. 2013.
7. A. Agarwal and A. K. Jagannatham, "Optimal adaptive modulation for QoS constrained wireless networks with renewable energy sources," *IEEE Wireless Communications Letters*, vol. 2, no. 1, pp. 78–81, Mar. 2013.
8. H. Liang, B. Choi, W. Zhuang, and X. Shen, "Optimizing the energy delivery via V2G systems based on stochastic inventory theory," *IEEE Transactions on Smart Grid*, to appear.

9. L. X. Cai, Y. Liu, T. H. Luan, X. Shen, J. Mark, and H. V. Poor, "Sustainability analysis and resource management for wireless mesh networks," *IEEE J. Selected Areas of Communications*, to appear.

10. S. Sarkar, M. H. R. Khouzani, and K. Kar, "Optimal routing and scheduling in multihop wireless renewable energy networks," *IEEE Transactions on Automatic Control*, vol. 58, no. 7, pp. 1792–1798, Jun. 2013.

11. T. Zhu, S. Xiao, Y. Ping, D. Towsley, and W. Gong, "A secure energy routing mechanism for sharing renewable energy in smart microgrid," in *IEEE Smart Grid Communications (SmartGridComm)*, Brussels, BE, 17–20 Oct. 2011, pp. 143–148.